Lady Leathernecks

The Enigma of Women in the United States Marine Corps

By Connie Brownson

NEW FORUMS

NEW FORUMS PRESS INC.

Published in the United States of America
by New Forums Press, Inc.1018 S. Lewis St.
Stillwater, OK 74074
www.newforums.com

Library of Congress Cataloging-in-Publication Data Pending

This book may be ordered in bulk quantities at discount from New Forums Press, Inc., P.O. Box 876, Stillwater, OK 74076 [Federal I.D. No. 73 1123239]. Printed in the United States of America.

ISBN 10: 1-58107-288-0
ISBN 13: 978-1-58107-288-4

.

To all the Marines I've loved,
female and male,
weak and strong,
"short" and "long,"
past, present, and future

Semper Fidelis!

Table of Contents

Preface

Operation Desert Shield/Storm

This work has been in progress, in my mind and on paper, since at least the spring of 1991 with the advent of Operation Desert Storm. An activated reservist myself, working twelve and fourteen hours a day leaving my toddler with a close friend and my active duty spouse in Drill Instructor School in San Diego, I looked around the Camp Lejeune Field House at the hundreds of faces of fellow activated reservist Marines, male and female, who had left homes, families, and jobs to fulfill their military obligation to guard our country. President George Bush's cause seemed righteous. The Marines were prepared. However, *everyone* within the rank structure, including the civilian leadership, was stymied by the sex/gender complexity of the activation and deployment. Basically, the female Marines, who had been "fully integrated" into their home reserve units, must stay behind because, legally, they could not deploy to a combat zone. So many questions arose: Where would we house the females at Lejeune? Should they continue to train with their units as they had in the past even though gaps would be left when they remained behind? What "jobs" would they do stateside when their units/male counterparts deployed? The logistics were complicated; the emotional aspects were heart-wrenching.

The All-Volunteer Force had been in place since the 1970s, including the integration of women. Did no one ever believe the day would come when units comprised of male and female Marines would actually need to deploy to defend national security? Contrary to the ROA National Security Report titled "Gulf War was a Test of Reserve Components and they Passed" by Honorable Stephen M. Duncan, then Assistant Secretary of Defense for Reserve Affairs, and other official reports, the mobilization of Marine Corps Reservists at MCB Camp Lejeune was controlled chaos. No one had a definitive plan. Marines adapt and overcome, of course, but the challenges encountered during Desert Shield/Storm should have been anticipated and appropriate operational policy and procedures established *before* the need arose and compromised successful deployment of American troops to war. Still, as politics would have it, the dog-and-pony-show floated across the sea and rolled across the sand.

No Single Explanation can be more than Partially Inclusive

When I returned to undergraduate study at Texas A&M University in 1993 (leaving with a shred of dignity after a brief academic tour there in 1983-84 before they could kick me out), I thought about my active duty Marine Corps experience. Through my coursework in sociology, I began to identify many aspects of that experience as emergent from interactions grounded in sociological theory. Clearly, a strong Durkheimian element presents, which identifies the separation of the sacred from the profane, the functionality and pragmatism of social systems, and, of course, the emphasis on ritual. Simmel's "Stranger" is relevant with military personnel existing in an extreme microcosm in a profession quite apart from mainstream America, as well as the military's compliant submission to, yet defense, of civilian government. Weber's influence, too, is evident in the various aspects of leadership ranging from charismatic to authoritative. Of course, the symbolic interactionists and exchange theorists *en masse* cannot be ignored as I witnessed the constant social exchange of tangibles and intangibles, particularly the use of emotions, favors, and sex as commodities.

I looked back on those dozen years of my Marine Corps life, and I truly believed that sociology held all of the theoretical and explanatory answers to accurately and wholly describe the phenomenon *if* I could just put it all together. I found, however, that sociology as a discipline is not enough to describe the complexity of the Marine Corps sex/gender relations phenomenon.

Another decade later, in graduate school at Texas State University while working full-time at The University of Texas at Austin, I finally identified the fundamental elements of the Marine Corps experience: physicality/biology, sexuality/gender, and power/submission manifested throughout social relations and military leadership. Sociology remained my prism for viewing the social world but there was something else, I believed, that was still hiding just out of sight. Perhaps, even, in plain sight, but my untrained or immature eye could not discern it. My sociological imagination ran wild while I worked for the Vice President for Research and took Master's classes at night. One day, a publication of Dr. David Buss, a prominent evolutionary psychologist on the UT Austin campus appeared, literally, in my mailbox as an entry in the Hamilton Book Author Awards program I administered: *Evolutionary Psychology: The New Science of the Mind*. Before I sent it out to the committee for review, I read it voraciously. I realized that Buss' theories and findings nicely supported my gut belief that human concepts of social organization actually emerge from something much more basic than the various sociological replication and conflict theories I had been taught. I became a believer that biological, evolutionary hardwiring was indeed at work as the basis for the social phenomena I had witnessed in the Marine Corps.

To that point, in all of my sociology classes, no professor has ever been able to answer my question, "*Why* do humans socially organize ourselves in the ways we do?" The responses were always anecdotal, and the discussions began with Durkheim and structural functionalism, or Compt, or Hegel, or Marx, or whomever. I was never satisfied with that. In the Enlightenment-driven, positivistic search for the "universal truth" about human social behavior, I think we have overlooked it because we are afraid of and appalled by it. In a positivistic world where all "truths" can be known and all "wrongs" righted, a more primeval "backward" view of human behavior equates to anarchy. Religion, philosophy, and numerous other disciplines turn the focus of "human nature" soundly away from a basis in biology. But, as I and others have found, this becomes a problem when we really have to get down to the more universal basics of "*Why?*" Social constructions appear to be relative but they also seem to emerge from universal biological imperatives.

Like David Buss, Matt Ridley, James Brain, and other similar researchers, I am no biological determinist. I disdain, for numerous reasons, the misuse of the term "natural selection" when referring to human evolution, primarily because "natural" implies that a result occurs without any human agency. Rather, we are sentient creatures who exercise free will. We culturally establish taboos, rules, and laws in order to protect the vulnerable, include the disenfranchised, etc. As a society, we laud altruism; however, we also create ingenious and horrific methods of intimidating, torturing, and destroying one another. We embrace and delight in power. We also cannot overlook the element of chance in everyday life. Sometimes the most socially proficient and prominent individuals are assassinated. Some of the most physically and reproductively fit human specimens die in accidents, such as automobile accidents, drownings, or plane crashes. Some of the most creative minds simply self-destruct.

Human existence is not as rational as we would love to believe. We embrace the rational because it is anticipated and should be controllable; it is "safe." But what if the irrational and the emotional are equally, if not more, prevalent in our everyday lives? Thus, Buss' theories and those of like-minded theorists provide insight into the *Why* of social behavior emergent from biology, and only from that point can sociologists proceed realistically to the *How*. Only through identifying and understanding the intersections of these theories, biological *and* sociological, can "human nature" be discussed in a realistic context in attempt to understand activities in day-to-day human existance, past, present, and even future.

Operation Enduring [Our] Freedom

Returning to the military context in 2015, although officially withdrawn, American troops are still in Iraq and Afghanistan hunting the "bad guys." The region remains in chaos. Humanitarian crises vie for media attention; terrorists strike with impunity. It's challenging to tell the "good guys" from the "bad guys." Similarly, sex/gender relations woes in the American military remain focused on the "bad *guys*." But who *are* they? Antiquated allegations of patriarchy and misogyny continue to drive the discourse about limitations on the total integration of women into the U.S. military. American women are "equal" and "free" to achieve whatever they want in life, yet sexual assault and rape capture headlines and the attention of government and citizenry. Perhaps, however, radical feminists' idealistic expectations are, in fact, limited, their focus is on the wrong enemy and their tactics are counterproductive to achieving their mission of women's acceptance by the organization and their colleagues.

In contrast to the militant tactics of radical feminists and "living legends" vignettes, this research presents female Marines' experiences, opinions, and suggestions for women's success in the USMC.

These women live the experience, providing nuanced, sometimes colorful, details of day-to-day military life that can never be captured by climate surveys and Likert scales. They share their mistakes, heartbreaks, successes, and love of God, Country, and Corps. They allow us to experience vicariously the enigma of being a female Marine. Not essentialist but pragmatic, their sense of kinship and equivalence provides a practical foundation for comprehending and, perhaps, "fixing" what allegedly is wrong with the U.S. military's sex/gender relations.

Acknowledgements

Although too numerous to list here, there are a few entities and individuals whose involvement must be recognized as directly instrumental to this work's completion. First, I thank The Marine Corps Heritage Foundation for its generous funding, but my gratitude runs much deeper than just for the money I received. The Foundation validated the importance of a project that I have been thinking about for over 20 years. This grant made me believe that someone would actually *care* to hear our stories and recognize how much we, as female Marines, are a vibrant part of America's military and social history as individuals and collectively.

Thanks, Mom, for everything! You did a great job as mother *and* father to me. I love you, Alexandra, Charley Elizabeth, and the wee one we have yet to meet. Be brave enough to follow your dreams, girls, but be wise enough to dream realistically.

Thank you to my Drill Instructors in WRTC, K Company, Platoon 6A, Senior Drill Instructor SSgt Michelle Holmquist, SSgt Deborah Phillips, SSgt Annie Reid, Sgt Anita Stargel, and Cpl Elizabeth Ayure, for starting me along this amazing path in January 1985. You made us march and sing to Billy Ocean's "Caribbean Queen," but I forgive you. As a young Marine, I made many of the mistakes you warned us about, but I hope that, ultimately, by bringing many of them to light here, I have validated your trust and confidence. Sergeant Major Dennis W. Reed, I thank you for your wisdom, leadership, and stoic good humor. You'll always be my hero. Thank you, Colonel James E. Morley, Major Timothy J. Fox, and LtCol William G. Wickun at MCRD San Diego, for inspiring me to return to college to complete my baccalaureate degree. Thank you, Captains Marty Korenek, Sam Clonts, and Jay Meynier, for your lessons in kinship and equivalency at MCB Camp Lejeune. I had no idea what I was learning from y'all at the time. We were just being Marines, which makes the experience that much more authentic, beautiful, and valuable.

I extend my sincere gratitude to SgtMaj Cherry McPherson, and Captains Teresa Ovalle, Charlotte Brock and Kate Hendricks for their personal and logistical support. I would like to also thank Sgt-

Maj Tamara Fode for her professionalism, logistical prowess, and genuine kindness, as well as that of the staff of the Marine Memorial Golf Course at Camp Pendleton. I also thank Maj Eric Dent of Headquarters, Marine Corps; Col James B. Seaton III, Commanding Officer of MCB Camp Pendleton, and LtCol Chris Hughes, I MEF PAO, for their personal acknowledgement of me as former Marine and social scientist, and their validation of the importance of this project even when others dismissed and dishonored me and my research.

I thank Vice President for Research, Dr. Juan Sanchez, and Drs. Diana Davis, Jeff Chipps Smith, Nassos Papalexandrou, and Katherine Arens for their sage advice, constructive criticism, and intellectual support of this work. Nassos, I especially thank *you* for telling me to not let anyone make me write the book they wished they could write, but to write my own! This is it; I really hope you like it. Thank you to Drs. David Crew (History), Victoria Bynum (History at Tx State), and Robert Kane (Philosophy) for inspiring me in the classroom. Although, perhaps not obvious, y'all are very much in the details of this project.

I owe a huge debt of gratitude to Dr. Patricia Shields at Texas State University and the editor of *Armed Forces & Society* for stimulating my intellectual curiosity and encouraging me to see past the words on the page and liberate the underlying concepts and associations in ways I never imagined I could. Also, I must thank Dr. James Burk, Department of Sociology, at Texas A&M University. I still have the books and notes from SOCI 313 in Spring 1993; check the Bibliography. These works continue to be relevant. Your course inspired me to be a military sociologist, even if it is happening later in life, part-time, and freelance. Thank you, Dr. Stejpan Meštrović for your edgy and spot-on irreverence; I really did do it for Durkheim. Thank you, Dr. John Tiefenbacher at Texas State University, for the faith you have placed in me as a scholar and the respect you have shown me as a colleague. You are the Best. Doctoral. Advisor. Ever. Thanks, also, to Doug Bynum, an incredibly supportive boss who understands and respects working one's way up in the world, and Brian McKay an all-around great guy and treasured co-worker even though he's a Wingnut. Thanks, also, to Drs. Eric Morrow and Kelly Lemmons for hiring me sight-unseen to teach Geography at Tarleton

State University. My word is my bond; appropriate results follow. Thank you, Milena Christopher, for compassion and wise counsel.

Lastly, I must thank The University of Texas at Austin's Departments of Sociology *and* History for denying me admission to their doctoral programs not just once, but *numerous,* times. I guess y'all were right! I never would have been content, happy, nor successful in your departments. Your rejection was truly a blessing. Instead of my intellect and passion being shackled to your coursework and research as a minion GRA, I devoted my time and energy to complete this incredible project and many others. *Memento mori.*

To all Marines everywhere: Stay motivated and *Semper Fidelis!*

> Connie Brownson
> May 10, 2015
> Rising Star, Texas

Chapter One

Introduction: The Fewer, the Prouder, the Female Marines

Situation

U.S. Army General William Thornson once said, "There are only two kinds of people that understand Marines: Marines and the enemy. Everyone else has a second-hand opinion." Physically, mentally, emotionally, and morally challenging, it is an experience unlike any other. It is an endeavor that few Americans, male or female, choose to undertake. Like the men joining, the Marine Corps requires young women to recite and truly embrace the credo, "I am an American, fighting in the forces which guard my country and our way of life. I am prepared to give my life in their defense," with the knowledge that the job requirements may also include violently taking the lives of other human beings.[1] The small number of females in this isolated, youth-centric, male-dominated, and extreme physical environment further complicates the occupational and social situation. While it is undeniable that women have entered into non-traditional occupations for centuries and, thus, is not unique, the experience of females in the Marine Corps is singular.[2]

Historical opinions about the presence and role of women in the military were largely dichotomous. The expectation in a few well-meaning camps has been that females should/would bring a "woman's touch" to warfighting/peacekeeping.[3] Others, however, have and continue to allege that the military, by its very nature as a war-fighting body, is an environment that must remain male-dominated because the inclusion of females weakens the infrastructure, jeopardizing every individual in a military unit on a micro-scale (e.g., the fire team) as well as national security on a macro-scale.[4] The contemporary reality of female incorporation into approximately 400 combat occupational specialties beginning in 2016 demands that these opinions and concerns be addressed in a non-emotional, practical, and insightful manner.[5]

Without first-hand knowledge of their experience, contemporary American awareness and opinions of female military service members emerge largely from the news media. From this, two very dramatic and opposite roles typically characterize the public's perception of military women. One role is the female embodiment of barbaric masculine attitudes and behaviors as perpetrators of high-profile abuse and torture crimes at Abu Ghraib.[6] Army Specialist Lynndie England characterized the embodiment of this role. Another extreme stereotype is the depiction of the female as perpetually frail, requiring protection and/or rescue by males. Army Private Jessica Lynch embodied this role. Both women featured prominently in the news headlines in the early days of the Iraq War. Victim status was also solidly reinforced by multiple murders of military women within an eight-month period in 2007-2008. These extreme "ideal types"[7] characterize the public's understanding of the experience of the American female in the military. Based in popular media constructs, stories, perceptions, and expectations emerge that titillate but do not accurately inform viewing and reading audiences. This further complicates understanding of the military experience and summarily discredits the efforts of the many and diverse women who live their military lives between these extremes.[8]

Similar to the media's influence on Americans' perception of military women, contemporary academic literature provides a biased radical feminist perspective that inhibits a pragmatic and thorough discussion of women's experiences in the military.[9] Within the organization and in the larger society, military females continue to be categorized and stigmatized.[10] "Patriarchy," identified overtly and subliminally,[11] has long dominated discussion as explanation for the perceived inequity in sex/gender relations in the U.S. military. This concept, even if an historical fact, no longer reflects reality. It is best simply and absolutely discarded.[12] Understanding gender relations in the military requires a thoughtful and cohesive analysis of issues related to women in the military from a solid scientific as well as social science framework, a synthesis of biological fact and sociological theory. Because both biological sex and the social construction of gender create stigma and the limitations to women who aspire to be equivalent service members, focus on only one side of the equation does not suffice. One aspect, whether sex or gender and the implications of each, simply cannot be separated from the other. [13] Queer theory, primarily defined as an attempt to undermine an overall discourse of sexual categorization and, more particularly, the limitations of the heterosexual-homosexual divide as an identity, occupies no place in this discussion. The sexualities of the participants in this research clearly remain constant and stable temporally and spatially, providing a cohesion unchallenged by the tenets of queer theory.[14]

This work endeavors to inform the scholarly literature as well as the lay reader. This demands a transdisciplinary methodology, analysis, interpretation, and presentation.[15] Filling a void in the literature and to inform the public, this qualitative research focuses on the United States Marine Corps's inclusion of women over the past 70 years. It identifies, through the innovative concepts of kinship and equivalency, how female Marines holistically reconcile their strengths and limitations within this unique social and occupational environment.[16] To date, no one has provided a researched and insightful portrayal of what it is to *be* a female Marine. They live every day in an environment of contradictory expectations, liberated to some degree by their military rank and its related social position, but still weighted down by arduous negative stereotypes.

This project endeavors to tell these stories accurately and fairly through the female voices of those who have earned the title "United States Marine," officer and enlisted. Using a transdisciplinary approach complementing theories primarily from evolutionary psychology and sociology, the analysis strives to explain why sex/gender disconnects persist in the Marine Corps. It also establishes an innovative and viable framework within which to discuss, evaluate, and potentially modify belief systems and behaviors to ensure personal and operational success of females in the U.S. military.

Mission: Thesis

I propose that the on-going "problem" of women in the military persists because of the unrealistic expectation that females are "equal" to males.[17] To address this, I assert that females and males are not equal, but "equivalent." Equivalency, as a concept, acknowledges without prejudice the differences of physical biology and socially constructed gender in sentient individuals enacting personal agency within environments of conditional interaction. Further, equivalency focuses on the kinship environment of the US military, similar to an extended family, emphasizing maximization of the contributing qualities an individual brings to the exchange. Those who are embraced as kin garner acceptance, protection, and even love. The concept of equivalency has the potential to neutralize physical disparity and emphasize cooperation and commitment to the common good.[18] This fits perfectly with Marine Corps leadership ideals, traits, and principles, which are not inherently gendered. Further, ancient Chinese philosopher Lao Tzu described leadership at its best when it imitates the most harmonious ways of nature, without harshness or aggressive struggle, and marked always by a gentleness that pulled subordinates naturally to their tasks.[19] My data, without question, support the concepts of kinship and equivalency in the U.S. Marine Corps. They demonstrate that individual female actors contribute to or undermine individual and mutual success using innate and learned characteristics, specifically the sex/gender they embody and employ in their lives, as female Marines and as women.

The purpose of this work is to provide insight into sex/gender relations in the Marine Corps since World War II. Sexuality and gender have significantly impacted females' potential for success since the inclusion of women in its ranks initially as ladies and today as warriors. Allegations of misogyny, radical feminists' demands for female "equality," and legal and moral mandates to eradicate sexual harassment, assault, and rape continue to tarnish the reputation of the U.S. military.[20] Informed by ill-considered assumptions, expectations, and allegations, the American public remains incredulous that negative stereotypes and sexual indiscretion, moral as well as criminal, have not been obliterated from the military.[21] In light of such scrutiny, sexual and gender expression by individuals in the military remains problematic. This research proves misogyny a myth. It showcases the overt and covert sexual power of women. Most importantly, it provides insight into the complexity of gender and sexual expression in the extreme environment of the Marine Corps. Finally, it reinforces the concepts of kinships and equivalency to augment existing ethical, operational, occupational, and legal systems to ensure the success of all individuals, male and female, as well as the warfighting organization.[22] Until women and men are freed from absolute sex/gender roles that imprison both, unresolved tension between the sexes will continue. To resolve, perhaps *both* sexes must revolt. Men are not the enemy. Women are not the enemy. To be anti-woman or anti-man is to be anti-life.[23]

Execution: Methodology

This research project is exploratory in nature. It began as an oral history project; I sought stories of how the experience of female Marines changed over time. The case study focuses solely on the experiences of female Marines, which is interesting and informative for a number of reasons. First, scant scholarly literature provides insight into their collective experience. Second, because of the extreme and isolated environment, they are captives (i.e., under legal contract) in their demanding profession and are geographically restricted by assignment. Third, the skewed ratio of males (93%) to females (7%) is less than half the overall percentage of females in the other U.S. military branches (~15%). Finally, in this overwhelmingly physical and youth-centric environment, the majority of Marines both male and female are in their mating years, ages 18 to 30, comprising 82.9% of the force, with an average age of 25. This is the lowest average age and four years younger than the other American military service branches.[24] The sample of female Marines interviewed for this project (n=179) is substantial. Four World War II interviewees participated as well as eight former female Marines. Thus, 167 participants were on active duty at the time of their interview.

The project investigated the female experience in the gender-skewed microcosm of U.S. Marine Corps, the toughest and most masculine of all military organizations.[25] The respondents self-selected; they *chose* to participate through snowball sampling because they felt they had a story to tell or insight to share. I did not seek out "superstars" or "groundbreakers," although a few female Marines who accomplished "firsts" appear in this work. For my purposes, however, I invited the rank-and-file, throughout the chain of command, and in all MOSs and geographic locations. Basically, any and all female Marines interested in the project who came forward to be interviewed were interviewed. Through this method, bias must be acknowledged. For example, females unsure or dissatisfied by their status in the organization or troubled about possible repercussions from their participation likely declined this opportunity. Additionally, the perceived sensitive nature of the subject may have restricted the sample size as well, which can be true when researching any exclusive group, especially those with highly charged values and opinions. Long interviews took place face-to-face at numerous locations, on and off Marine Corps bases across the United States, as well as via telephone and e-mail. Conversations with female Marines stationed around the globe from Iraq to Iwakuni avalanched as one Marine spoke to another and then to another, spreading the excitement that someone wanted to listen to their stories.

Of significance to this project is the fact that female Marines are indoctrinated with the leadership principle, "Know yourself and seek self-improvement." In contrast to Kier's assertion that organizational actors are usually unaware of beliefs and values that guide their behavior[26], female Marines are an extremely self-aware group and, as one will read, are not shy about sharing their thoughts, beliefs, and experiences. Although self-selected though snowball sampling, the interviewees provide a strong and consistent voice which led to a grounded theory analysis. One described the interview process as "therapeutic." Numerous others expressed gratitude that they, as female Marines, would finally have a public voice. It is because of this that the interview quotes herein remain in their vernacular. The quotations selected for inclusion herein are not aberrations, but those that most succinctly enunciate the phenomenon discussed. All female Marines described and quoted here are real; all but one requested that their real identities be used. However, since time has passed since they offered exposure of their identities for public scrutiny and they are now to the winds, pseudonyms have been used herein to attempt confidentiality in respect for their current circumstances.

Rather than retooling previous assumptions and hypotheses that exist in the literature, the data analysis investigated linkages of data to theory primarily from evolutionary psychology and sociology. It identifies and postulates how fundamental human sexual behavior emerges from the evolutionary hard-wiring in the brain and couples with social constructions of gender to create defined but malleable interactions and symbiotic relationships between the sexes. This complex interaction of biology and sociality manifests centrally in the contemporary debate about females in the U.S. military particularly because of the extreme isolation, physicality, youth-centric, and gendered nature of the profession. I acknowledge that the voices that speak here can be interpreted in many ways. I offer only one of many possible interpretations in the hope that discourse on sex/gender relations in the military may be expanded beyond the currently dominant radical feminist dogma.

Evolving Generations of Female Marines

In contrast to Judith Hicks Stiehm's cohorts,[27] I identify inductively through data collection three, and an emerging fourth, generations of female Marines. The first group, the "Free a Man to Fight" generation, includes women who were in the Marine Corps during World War II through the Vietnam era. Whether out of patriotism or boredom, these women entered a military environment that mandated and monitored strict separation of the sexes, personally and professionally. Physical and military standards for the two sexes were incomparable. Women did not "lead" men; they hosted tea parties. They filled limited billets approved for their sex/gender. Although they wore the "Marine" title, they lived and worked largely separated from the male fighting and support forces within an organizational/management structure consisting almost exclusively of other females.[28]

The second generation, the "All Volunteer Force" generation, includes Women Marines (WM's) from roughly 1973 through 1991. Whether they considered themselves feminists or not, these trail-blazing women embodied the feminist ideals of the time and entered a man's world to do a man's job for a man's pay. What they encountered was a mixture of acceptance, confusion, and sometimes outright hostility. Occupations were still strictly limited, and males' and females' physical and military standards were miles, literally, apart. Both sexes negotiated the uncertain terrain of evolving sex/gender roles and expectations. For example, in recruit training, females were instructed in proper make-up application and were required to wear at least eye shadow and lipstick. In the late 1980s, females transitioned from "fam" (i.e., familiarization) firing the M-16 rifle to actual qualification and began to drill with rifles and perform the Manual of Arms. Although substantive advances occurred, there was never the expectation or anticipation that females in the Marine Corps would be considered "warriors" during this period. This generation "ends" with the Tailhook scandal and the launch of Desert Shield/Storm. Both events illuminated, professionally and organizationally, the unique sex-based challenges faced by females in the military and demanded an evaluation of gender relations in the military both in garrison and the field.

The third generation is the "Post-Tailhook" generation that served in an American military clearly marketed to the public as misogynist, openly hostile, and dangerous to females.[29] The Navy's 1991 Tailhook and the Army's 1996 Aberdeen Proving Ground scandals defined this period, which spans roughly 1991 to 2003 with President Bush's launch of Operation Iraqi Freedom. The female Marines of this generation lived and worked in an environment where their standards historically and organizationally most closely matched those of their male counterparts and actually approach equity. They are expected to be "Marines," no longer designated as "Women Marines" or just "WM's." All military occupational specialties, except those expressly designated as "combat" by Congress, are open to them. Even that presumably sharp line in policy blurs in reality as evidenced in Iraq and Afghanistan where females serve and die in combat actions.[30] Of note also is the fact that the modern military is an environment with its own brand of within-gender dichotomy: those committed females who are successful versus those who are not who thereby bring discredit to the entire group. Also during this time, an unfortunate irony emerged. In response to the Tailhook scandal, the Secretary of the Navy announced a "zero tolerance" policy on sexual assault by members of the naval services, which includes the Marine Corps.[31] At the point when female Marines are closest to their male counterparts in terms of occupational equity, the advances made were eclipsed by their categorization as victims of sexual harassment, abuse, assault, and rape. The question to be answered now is: How can a female be a competent military leader if her sex renders her vulnerable?

The current, emerging generation can aptly be labeled "Women as Warriors." This generation comprises women who historically have seen physical fitness standards, training, and military leadership expectations most equal between the sexes. Occupationally, they have the opportunity to enter combat training and will experience the entry of women into all combat MOSs previously restricted to males. Women who earn the right will be *bona fide* leaders of men in the most extreme of human experiences: military combat. Military and societal accolades will follow, but the status also encumbers great responsibility at home as well as on the battlefield. As military leaders in garrison, they will negotiate increasingly rigorous and complex sexual interaction regulations for themselves and their troops. For example, recent legislation such as the expansion if Title IX[32] and California's "Yes Means Yes"[33] will have profound implications for the U.S. military. Although well-meaning, these laws reify females' sociosexual inferiority and vulnerability. Further, they place the sociosexual restraint onus wholly upon males. As expressed in my data, this hardly seems equitable and complicates sex/gender relations in the Marine Corps.

As would be expected, these categories include individuals whose careers overlap the chronological boundaries established, prevailing social attitudes span these generations, and institutional policies take time to implement and their results identified and actually measured. That said, however, they serve as valid units of categorization for this discussion. To reiterate, the goal is to construct an innovative and purposeful framework from which to investigate and address ongoing, challenging sex/gender relations in the U.S. military.

In the spirit of Kingsley Browne,[34] this work wholly rejects political correctness that historically has stymied the acquisition and assimilation of knowledge of this subject into a cohesive, descriptive, and explanatory body. This work presents results of an independent, freelance investigation and analysis. As sociologist Anthony King queries, the question now is whether the concept of "equivalency" adequately captures the scale of the current changes substantially enough to facilitate full accession of women into ground combat roles.[35]

Administration & Logistics: Literature Review

One of the first intellectual attempts to truly understand gender differences in the military from a researched theoretical perspective was undertaken by Dr. Christine Williams in 1989 who published

Gender Differences at Work. In this seminal study drawn from her doctoral dissertation, she compares the gender construction/maintenance experiences of male nurses to those of female Marines. While Dr. Williams makes many excellent observations, she fundamentally misses the essence of what it is to be a *female* Marine primarily because of the limited scope of the two largest groups in her pool of military interviewees.[36] Her first mistake (acknowledged in the book's Forward) was to include a large number of recruits (n=21) in her interview pool. This is a critical error because her discussion focuses on gender differences *at work* and, due to the separation of male and female recruits during initial training, the women interviewed had not even entered the Marine Corps "workforce" yet. Female recruits are in an environment of extreme isolation where their activity for almost every hour of their day is dictated to them. Although expected to perform traditionally masculine tasks (e.g., close order drill with rifles, throwing grenades, "humping," etc.), in the absence of the mere presence of the other sex, discussion of the recruits' "construction" of gender is moot. In addition, the inclusion of a second large group, female drill instructors in her sample (n=14), skews Williams' data further simply because of the elite nature of that billet. Drill Instructors are the personification of the Marine Corps' leadership traits and principles and physical fitness – they are *not* the ordinary military workforce. The experience of female Marines from recruit training through separation from active duty is much more rich and varied than the insights these two primary groups suggest.

Beyond Williams' seminal work, "women in the military" publications since the Gulf War, such as Erin Solaro's *Women in the Line of Fire: What You Should Know About Women in the Military* (2006) and *One of the Guys: Women as Aggressors and Torturers* (2006), edited by Tara McKelvey, are both written by journalists as a personal account and a collection of essays, respectively, not as scholarly discussions of women in the modern U.S. military. Although quite different in format, they both exhibit biased portrayals of the military services rife with violent men and victimized women. The insights advanced in these publications are either from the author's personal, external perceptions of military situations or gleaned from testimonies of the military men they encountered. They possess extremely loose "validity" as experiential literature while not possessing the methodological rigor of a true social scientific investigation into the social phenomenon, which connects data to theory.

In contrast, Kayla Williams' 2005 *Love my Rifle More than You*, a personal narrative of her experience in the Army, although not corroborated by empirical evidence, frankly asserts: "Sex is the key to any woman soldiers' experiences in the American military."[37] Similarly, Melissa Herbert's *Camouflage isn't Only for Combat*, in 1998 acknowledges the U.S. military as a workplace, identifying the issue of sexual expression and enactment as disruptive. She states that sexuality in the workplace is seen not as an inherent part of the social organization of work, but as an intruder into a space perceived as both rational and asexual.[38] Rosabeth Moss Kanter's influential concept of tokenism captures this phenomenon in a strictly static way as she describes "auxiliary traits" within an organization. Specifically, tokens differ not in ability to do a task or in acceptance of work norms, but in terms of secondary and informal assumptions of ascribed characteristics that differ from the majority group.[39] In the case of the military, it is a female's sex/gender enactment that generates disruptive assumptions and expectations.

Conservative views, such as Brian Mitchell's *Women in the Military: Flirting with Disaster* (1998), raise many excellent points about the difficult integration of women into the U.S. military and the service academies. Mitchell acknowledges many larger cultural explanations for the seemingly revolutionary changes taking place in these institutions. The importance of Mitchell's work to this project is summarized in his statement regarding the grievous error of assuming inherent guilt or innocence of either sex in the military environment:

> Once again the principal victims were not narrow-minded traditionalists opposed viscerally and ideologically to the advancement of women, but forward-thinking, fair-minded men willing to give a competent woman a chance. They just didn't understand how things work in the real world, how powerful sex is, how craven some men are, and how spiteful and manipulative some women can be.[40]

In a similar mentality, military historian Martin van Creveld asserts, "…judging by their respective relationship to war, the social roles of men and women are not nearly as flexible as some people believe or want others to believe."[41] Both of these writers speak to the relative power of each sex in the military environment and in relation to warfighting. Contrary to van Creveld's assertion of inflexibility, relativity innately implies flexibility. Male-female physical and social relationships are not static, but coevolutionary.[42]

Feminist Linda Bird Francke's *Ground Zero: The Gender Wars in the Military* (1997) describes the challenges of mixing the sexes in the military environment, physically as well as socially. However, she summarizes the futility of the situation by stating, "The resistance to women will not go away because it can't….the dynamic of the white male culture that draws men to it depend on exclusivity." Blaming almost wholly the societal aspect of the dilemma, she further concludes, "Instead of drawing the genders together, the dynamics at work in the military culture often force them apart… Judged against the majority male model, women have to be the Wrong Stuff…The cultural wars will never end."[43] It is interesting that Francke employs the term "cultural wars" to describe division between males and females of exactly the same culture; in this case it is the human, military, and American culture. Her assertion of division between the sexes in the military remains justified today by events on military installations and in the news feeds. Based upon my data, I allege that it is not the males' motivation for exclusivity driving division. In fact, the opposite occurs. For example, female Marines interviewed express frustration over systemic "favoritism." SSgt India captures this conundrum, stating:

> As long as the Marine Corps keeps making allowances for females, the leadership continues to drive a wedge between the sexes. When the system divides us, that's when the resentment begins. We aren't asking for anything other than what the men already have. The system is giving the females things and what are we gonna say, "No?" We're screwed any way we respond. We're either a pain in the ass because they think we need to be "accommodated" or we're ungrateful. We lose either way. These are just non-issues for the males.[44]

Stephanie Gutmann, too, identifies bias brought into the institutional investigation of gender relations in the military. In *The Kinder, Gentler Military: How Political Correctness Affects our Ability to Win Wars* (2000), she states, "Year after year, since the mideighties, they [the military establishment] have hired the people who most distrust military culture and find sexual harassment around every corner; they publish studies portraying military life as a veritable hell on earth for women in the services...".[45]

Recent works edited by Cohn (2013) and penned by Sjoberg (2014) adopt a refreshingly comprehensive perspective of men's and women's investment in the process of war. For example, Sjoberg argues that seeing war as a system rather than an event and seeing war as separable from the politics (and violence) of everyday life is critical to mainstreaming gender in thinking about war and conflict.[46] Although a commendable effort, she fails to acknowledge and incorporate into her argument the *complicity* of women in these processes and their exercise of personal power to disrupt the organization and its mission as evidenced by my data.[47] Similarly, Cohn's edited volume strives to focus on the "gendered power relations, in all of their complexity and multidimensionality, which are at the heart of women's vulnerability in war."[48] Although Cohn rejects the false dichotomy of Woman as either victim or agent and its implications for policy, the essays included provide only snapshots of women's experiences, thus limiting understanding of a potentially larger phenomenon.[49] Women are not portrayed as individuals, but as women, a collective. History indicates that this undermines women both as individuals and collectively. This mentality manifests the same as patriarchy, but now advocated by radical feminists.

By advocating an us-versus-them mentality, both through the archaic benevolent patriarchal and now the radical feminists' reverse-patriarchal mentality, the "findings" advanced by the authors and editors mentioned here steal from military women cognizance of their sexuality and gender, components of the self that remain integral to their personal identities. This, in turn, quashes their individual agency to conduct themselves as responsible, or even irresponsible if they so choose, sexual and gendered beings. Thus, by consistently casting female service members as victims or vixens, authors such as these exercise literary theft of their sexual agency and its associated social agency. This theft affects *all* women in the military past, present, and future; thus, the negative stereotypes propagate. This theft of females' identity and agency will continue until an accurate and equitable mosaic of women in the military is presented that addresses the fundamental biological and social intersections of male and female interactions in the military environment. Until that time, conceptually joining the concepts of *woman* and *warrior*, and *female* and *Marine* defies any rational cognitive basis. An individual simply cannot be a successful military leader and combatant if s/he is a victim.

This project introduces a theoretical perspective grounded solidly in evolutionary psychology coupled with sociological structural functionalism. This synthesis of theories and data *incorporating* biological and social scientific perspectives provides the "missing link" between the dichotomous arguments that: 1) females, being equal to males, are perfectly capable of performing as warriors, and 2) females are not biologically or socially suited to the task of killing and war-fighting. This is accomplished by combining knowledge of human physiology and compelling theories of brain "hard-wiring" over the course of evolution[50] with a cultural preponderance by humans to maintain the biological and social status quo.[51] The data support the male-female kinship and equivalency concepts that facilitate the success of females in the Marine Corps. They further provide evidence of what a gender-integrated American warfighting organization could be.[52]

Command & Signals: Utilization of Theory

Proposed here is not a grand theory to explain the biological evolution and social coevolution of men and women over the course of humankind's existence. Instead, a few reasonable interdisciplinary associations plausibly explain why females in the United States Marine Corps continue to face many of the challenges and stereotypes that "women's liberation" was supposed to eradicate.[53] It reframes our awareness and understanding of how physicality, gender enactment, and institutional requirements frame the experience of women in the military.

Intersecting theories gleaned primarily from evolutionary psychology and sociology ground this discussion. Evolutionary psychologist David Buss asserts that culture and consciousness were presumed to free us from evolutionary forces.[54] Recognizing the conflict between a variety of "active forces," including potentially destructive, unrestrained competition for mates and the necessity of social rules in which reproduction could "appropriately" be pursued, sociologist Emile Durkheim also writes, "We cannot pursue moral ends without causing a split within ourselves, without offending the instincts and the penchants that are the most deeply rooted in our bodies."[55] In this, Durkheim may have been the first sociobiologist. He recognized that individual restraint required "social facts" to impose the will of the whole onto the will of the individual, which again describes the reciprocity between human biology and human society.

Recognizing that human society consists of a variety of exchange relationships between and among individuals, sociologist Peter Blau[56] advances the concept that social attraction is the force that induces all human beings to establish social associations on their own initiative. On the most basic social level, one individual may possess and/or command a commodity or service desired by another individual. Anyone who commands the sought-after commodity or service attains power over any other who desires it by withholding the desired item contingent upon satisfaction of his or her need. Sharing or relinquishing the desired item remains with the possessor. Within overlapping biological and social realms, per Buss, Durkheim, and Blau, a woman's sexuality is easily commodified. Ironically, rather than the woman herself maintaining power over her sexual commodity, in many societies a female's sex remains the property of her mate. If she is unmated, it belongs to her male kin for protection until it is transferred to another approved male.[57] In theory, however, if a woman could, within her own capacities, wholly manage her physical and sexual capacities against encroaching males, the need for male, or by extension legal "protection," would not be necessary.

To follow through with this line of thought, in the absence of any real proof of either biological or social determinism and domination and recognizing external, limiting forces on individual behavior, one may reasonably side with sociologist Max Weber, who asserted, "…every genuine form of domination implies a minimum of voluntary compliance, that is, an *interest* (based on ulterior motives or genuine acceptance) in obedience."[58] [Emphasis is in the original.] If Weber's assertion is accepted, one must assume that complicity in one's own subjugation through voluntary moral agency is not unlikely. Buss states from an evolutionary perspective, "Members of each sex, in essence, become willing victims to the whims and desires of the opposite sex."[59]

It is not surprising, then, that society has adapted social practices over the course of history to ensure at least some practical forms of domination and submission. This does not mean, however, that the process of adaptation completely overcomes basic biological imperatives or the enactment of free will and personal agency or that, once accomplished, any adaptation will be stable or tend toward universality. This merely means that the human experience is a give-and-take, with some forces being stronger at some times than others, potentially equally valuable, and involving a myriad of actions, reactions, and chance occurrences.[60] Founded on this premise, this research provides credible evidence of a trajectory from basic human biology to social agency within organizations and, ultimately, frames the individual and collective experiences of females as United State Marines.

The Call to Battle

Gender optics is the meaningful way by which we view and structure our relationship with an individual defined by his/her sex/gender. These gender optics must be calibrated to view the U.S. Marine Corps microcosm not through our preconceived expectations, but with an open mind to recognize the necessity for management of real human bodies performing physical and potentially deadly tasks within a conservative institutional framework. Relevant U.S. Marine Corps institutional attributes are: 1) *isolation* both geographically and ideologically from the larger American society; 2) *sex-skewed* demographic with 7% of the force comprised of females; 3) *physicality* that focuses on the physical body and its capacity for strength, endurance, and performance; and 4) *youth-centric* that situates individuals who are in their mating years and developing emotional maturity while situated in positions of authority over others and often in extreme situations, such as warfighting.

Philosopher and feminist theorist Marilyn Frye asserted that the word "virgin" did not originally mean a woman whose vagina remained unpenetrated by a male's penis, but a free woman, one not betrothed, not married, not bound to, not possessed by any man. It meant a female who is sexually, and hence socially, her own person.[61] In the military environment biological, gendered, and occupational strengths and limitations directly impact personal performance, acceptance, and success.[62] The Marine Corps organization manages or, some may argue, mismanages what other occupational institutions treat as *private* (e.g., promiscuity, pregnancy, fraternization) as part of its *public* (e.g., mental and physical strength, operational effectiveness) mission. Female Marines, Frye's Willful Virgins, are enigmatic for this very reason. Perhaps it is the female Marines' sense of independence, sense of self, potential for personal empowerment, and the *opportunity* to display and enact her physicality and prowess that generates confusion and discomfort in her occupational and social circles and, thereby, in American society at large. Perhaps it is her [mis]use of sexuality and/or femininity to garner acceptance or favors in a sex-skewed environment or to reject such opportunities to make her own brand of success. In this murky space of negotiating "appropriate" sex and gender roles, *both* male *and* female Marines struggle to amicably co-exist, while also remaining the finest fighting force in the world. The American public and government officials remain simultaneously mystified and polarized.

This research reflects female Marines' acceptance as well as rejection of prescribed social and professional roles and the corresponding acceptance or rejection by their Marine colleagues. The concepts of kinship and equivalency encompass these relationships without imposing the expectation that anyone, male or female, must transcend their biological or social nature to contribute in a relevant manner to the community's, individuals', and mission's success.[63] The data further imply that from kinship and equivalency, empathy and camaraderie develop. The findings are intriguing and have great potential for profound institutional and societal impacts.[64]

Endnotes

1 Elizabeth M. Culver, "Women in the Service," *Annals of the American Academy of Political and Social Science*, 229, The American Family in World War II (September 1943): 67.

2 Connie Brownson, "The Battle for Equivalency: Female US Marines Discuss Sexuality, Physical Fitness, and Military Leadership," *Armed Forces & Society* doi:10.1177/0095327X14523957 (March 2014): 16.

3 Carol Cohn, "Feminist Peacekeeping," in *The Women's Review of Books* 21, no. 5 Women, War and Peace (February 2004): 8.

4 There are too many to list. Referenced in this work, specifically, are Kingsley Browne, *Co-ed Combat: The New Evidence That Women Shouldn't Fight the Nation's Wars* (New York: Sentinel, 2007); Robert L. Maginnis, *Deadly*

Consequences: How Cowards are Pushing Women intoCombat (Washington, DC: Regnery Publishing, Inc., 2013); Brian Mitchell, *Women in the Military: Flirting with Disaster* (Washington, DC: Regnery Publishing, Inc., 1998); and Martin van Creveld, *Men, Women and War* (London: Cassell & Co., 2001).

5 Hope Hodge Seck, "Marines Delay Female Pullup Requirement Again, this Time Until the End of 2015," *Marine Corps Times*, July 3, 2014, http://www.marinecorpstimes.com/article/20140703/NEWS/307030068/Marines-delay-female-pullup-requirement-again-time-until-end-2015.

6 Stjepan G. Meštrović, *The Trails of Abu Ghraib: An Expert Witness Account of Shame and Honor* (Boulder, CO: Paradigm Publishers, 2007).

7 Hans Gerth and C. Wright Mills, trans. and eds., *From Max Weber: Essays in Sociology* (New York: Oxford University Press, 1946).

8 Connie Brownson, "Battle for Equivalency," 6.

9 Ibid., 4.

10 Ibid., 14.

11 Anthony C. King, "Women Warriors: Female Accession to Ground Combat" *Armed Forces & Society* doi: 10.1177/0095327X14532913 (May 2014): 2.

12 Connie Brownson, "Rejecting Patriarchy for Equivalence in the U.S. Military: A Response to Anthony King's 'Women Warriors: Female Accession to Ground Combat'" *Armed Forces & Society* DOI: 10.1177/0095327X14547807 (August 2014): 2-3.

13 Ibid., 3.

14 Tim Edwards, "Queer Fears: Against the Cultural Turn," *Sexualities* (1998) vol. 1 (4): 472.

15 Bärbel Tress, Gunther Tress, and Gary Frey, eds., "Defining Concepts and the Process of

Knowledge Production in Integrative Research," 13-26, *From Landscape Research to*

Landscape Planning: Aspects of Integration, Education, and Application (Wageningen

UR Frontis Series (Book 12) The Netherlands: Springer, 2005).

16 Brownson, "Battle for Equivalency," 3.

17 Ibid., 4.

18 Ibid., 3.

19 Malham M. Wakin, author and ed., "The Ethics of Leadership II" in *War, Morality, and the Military Profession* (Boulder: Westview Press, 1986): 201.

20 Ruth Rosen, "The Invisible War Against Rape in the U.S. Military, *HNN History News Network*, March 24, 2014, http://hnn.us/article/155049.

21 Jeanne Holm, *Women in the Military: An Unfinished Revolution* (Novato, CA: Presidio Press, 1982).

22 Brownson, "Battle for Equivalency," 2.

23 Eleanor Brantley Schwartz and James J. Rago, Jr., "Beyond Tokenism: Women as True Corporate Peers," *Business Horizons* 16, no. 6 (December 1973): 75.

24 Demographics of Active Duty U.S. Military, http://www.statisticbrain.com/demographics-of-active-duty-u-s-military/.

25 King, "Women Warriors," 2.

26 Elizabeth Kier, "Discrimination and Military Cohesion: An Organizational Perspective," in *Beyond Zero Tolerance: Discrimination in Military Culture*, ed, Mary Fainsod Katzenstein and Judith Perry (Lanham, MD: 1999): 47.

27 Judith Hicks Stiehm, "The Generations of U.S. Enlisted Women" *Signs* 11, no. 1 (Autumn 1985): 174-5.

28 Mary V. Stremlow, *A History of the Women Marines, 1946-1977,* prepared for the History and Museums Division, Headquarters, U.S. Marine Corps, (Washington, DC, 1986).

29 Patricia M. Shields, "Sex Roles in the Military," in *The Military-More than Just a Job?*, eds. Charles C. Moskos and Frank R. Wood (McLean, VA: Pergamon-Brassey's International Defense Publishers, Inc., 1988): 99-114.

30 "Female Troops in Iraq Exposed to Combat," *CNN.com*, World, June 28, 2005, http://www.cnn.com /2005/WORLD/meast /06/25/women.combat.

31 Michael Winerip, "Revisiting the Military's Tailhook Scandal" *The New York Times*, Retro Report, May 13, 2013, http://www.nytimes.com/2013/05/13/booming/revisiting-the-militarys-tailhook-scandal-video.html?_r=0.

32 David S. Cohen, "Title IX: Beyond Equal Protection," *Harvard Journal of Law and Gender* 28, no. 2 (2005).

33 California Lawmakers Pass 'Yes means Yes' Campus Sexual-assault Bill, *Los Angeles Times*, Local/L.A. Now, August 29, 2014, http://www.latimes.com/local/lanow/la-me-ln-california-yes-means-yes-sexual-assault-bill-20140829-story.html.

34 Browne, *Co-ed Combat*, 182.

35 King, "Women Warriors," 7.

36 Christine L. Williams, *Gender Differences at Work: Women and Men in Nontraditional Occupations* (Berkeley: University of California Press, 1989): 145.

37 Kayla Williams with Michael E. Staub, *Love My Rifle More than You: Young and Female in the U.S. Army* (New York: W.W. Norton, 2005).

38 Melissa S. Herbert, *Camouflage Isn't Only for Combat* (New York: New York University Press, 1998), 63.

39 Rosabeth Moss Kanter, "Some Effects of Proportions on Group Life: Skewed Sex Ratios and Responses to Token Women," *American Journal of Sociology* 82, no. 5 (March 1997): 968.

40 Mitchell, *Women in the Military*, 331.

41 van Creveld, *Men, Women and War*, 169.

42 David Buss, *"The Evolution of Desire: Strategies of Human Mating* (New York: Basic Books, 2003): 209-214.

43 Linda Bird Francke, *Ground Zero: The Gender Wars in the Military* (New York: Simon & Schuster, 1997), 260.

44 Brownson, "Battle for Equivalency," 15.

45 Stephanie Gutmann, *The Kinder, Gentler Military: How Political Correctness Affects our Ability to Win Wars* (San Francisco: Encounter Books, 2000), 213.

46 Laura Sjoberg, *Gender, War & Conflict* (Malden, MA: 2014): 143.

47 Brownson, "Rejecting Patriarchy," 2.

48 Carol Cohn, "Women and Wars: Toward a Conceptual Framework," in *Women & Wars*, ed. Carol Cohn (Cambridge, UK: Polity Press, 2013): 29.

49 Ibid., 31-32.

50 See Matt Ridley, *The Red Queen: Sex and the Evolution of Human Nature* (New York: Penguin Putnam, 1993); and David M. Buss, *The Evolution of Desire*, and *The Murderer Next Door: Why the Mind is Designed to Kill* (New York: Penguin, 2005).

51 See Thorstein Veblen, *The Theory of the Leisure Class* (New York: Penguin Books, U.S.A., Inc., 1899); Peter M. Blau, *Exchange and Power in Social Life* (New York: John Wiley & Sons, Inc., 1964); and Robert N. Bellah, ed., "The Dualism of Human Nature," in *Emile Durkheim: On Morality and Society: Selected Writings*. (Chicago: University of Chicago Press, 1975).

52 Brownson, "Battle for Equivalency," 21.

53 Jon Weinstein "Women's History Month: Leading Author in Women's Liberation Movement Continues Fight for Equality" *Time Warner Cable News NY1*, March 18, 2014, http://www.ny1.com/content/news/178878/women-s-history-month--leading-author-in-women-s-liberation-movement-continues-fight-for-equality.

54 Buss, *Evolution of Desire*, 3.

55 Emile Durkheim, "The Dualism of Human Nature," in *Emile Durkheim: On Morality and Society* (Chicago: University of Chicago Press, 1973), 152.

56 Peter M. Blau, *Exchange and Power in Social Life* (New York: John Wiley & Sons, Inc., 1964).

57 Buss, *Evolution of Desire,* 211-214.

58 Max Weber, "The Types of Legitimate Domination," in *Social Theory: The Multicultural & Classic Readings*, ed. Charles Lemert (Boulder, CO: Westview Press, 1993), 122.

59 Buss, *The Murderer Next Door*, 53.

60 Brownson, "Rejecting Patriarchy" 3.

61 Marilyn Frye, *Willful Virgin: Essays in Feminism, 1976-1992* (Freedom, CA: The Crossing Press, 1992).

62 Brownson, "Battle for Equivalency," 18.

63 Brownson, "Battle for Equivalency," 20.

64 King, "Women Warriors," 7.

Chapter Two

Infiltrating the Band of Brothers

Preeminent military sociologist Charles Moskos asserted that the military's genetic self-image is that of a specialist in violence, ready for combat.[1] Marines are distinct even within that separate world of the U.S. military. Theirs is a culture apart. Their values and assumptions shape the Corps' members; it is what binds them together. Theirs is the richest culture: formalistic, insular, elitist, with a deep anchor in their own history and mythology.[2] Perhaps seduced by the promise of adventure and camaraderie and posters and billboards of handsome, virile young men sporting high-and-tights, young American women arrive at Marine Corps recruiting offices nationwide from a vast array of ethnicities, socioeconomic backgrounds, and education levels. They possess a gamut of life and work skills as well as body types, athleticism, and confidence. They also arrive with a vast array of needs and expectations, but what do they seek? Author Brian Mitchell observes:

> …self-interest was the best focus for [recruitment of] military women. They discovered a new-found "right to serve." No longer were they merely support troops, freeing a man to fight. Now they had equal status, equal advancement, and equal benefits, and they were moving in large numbers into previously all-male units and specialties.[3]

Mitchell casts female recruits and officer candidates in an unfavorable light, implying that military women join selfishly, with less "noble" intentions than their male counterparts. The reasons cited by females for joining the Marine Corps, however, are no more self-serving than those of males. All of the female Marines interviewed expressed with obvious pride a distinct awareness of the unconventionality of their decision to join the Marines. For some, it was a calculated decision; for others, nothing more than a whim.

Motives prompting military women's service were of interest

even in 1943. Navy LTJG Elizabeth Culver wrote of the women serving at a pivotal period in WWII, "In every case, there is a very real, though frequently inarticulate, desire to do one's patriotic duty... there was the hope for excitement, the possibility of seeing action and far places..."[4] Similar to women's experiences during American Westward expansion, a mere 50 years prior, historians have found that women were active, often willing, participants in westward migration and settlement. In their journals and letters, many westering women expressed joyously their anticipation of "seeing the elephant," experiencing events, people [Native Americans], and locales they had never experienced before. Myers states, "Pioneer women were neither the sunbonnet saints of traditional literature nor the exploited drudges of the new feminist studies."[5] The obvious sociological dichotomy once again noted, the female Marine experience is not dissimilar.

The Postemotional Military

The hallmark of the Modern military was that of an institution legitimized in terms of values and norms based on a purpose transcending individual self-interest in favor of a presumed higher good. Members of the American military were often seen as following a calling captured in words like, "duty, honor, and country." Moskos' prophecy of the Postmodern military, however, envisions a foundation of identity politics based on ethnicity, gender, and sexual orientation.[6] Female Marines interviewed for this research and the organization they describe exhibit this transition from the Modern to the Postmodern American military. Specifically, sociologist Stjepan Meštrović's concept of postemotionalism, as a component of the Postmodern military, provides insight into this phenomenon. He defines postemotionalism as a state in which, "...synthetic, quasi-emotions become the basis for widespread manipulation by self, others, and the culture industry as a whole."[7] The epoch Meštrović analyzed is roughly captured within events of the 1960s and early 1970s, including culture-testing events such as the assassination of JFK, the Hippie movement, race riots, through the debacle of Vietnam. Interestingly, this profound cultural shift from inner- to other-directed corresponds directly with the implementation of the All-Volunteer Force.

Meštrović further identifies this shift of privilege in social organization as historical-generational, stating, "The other-directed focus on tolerance blossomed into the full-blown and highly organized cults of multiculturalism and political correctness."[8] In regards to the military, he draws upon Reisman's theory of inner- versus other-directedness:

> …in the past…tradition-directed and inner-directed soldiers held some firm values for which they were willing to make sacrifices, including the ultimate sacrifice. But other-directed individuals – the soldiers who are directly involved as well as the population at home who participate vicariously – find no transcendent value worthy of sacrifice. In the end, the only value left is survival of the self -- a conflicted position for a military *force*.[9]

When an individual's needs and wants are placed above social rules and expectations, and *all* individuals' needs and wants are considered equally relevant, an environment of coerced, not genuine, respect is created. Thus, in contrast to Mitchell's assertion that "feminization" has caused the demise of the U.S. military,[10] Meštrović's postemotionalism, not femininism, has infiltrated the military from American society to the detriment of the organization and national defense. Pervasive now are other-directed values (e.g., the rejection of a collective culture, elevation of the self above the group, and the erosion of empathy, loyalty, and etiquette, etc.). These concepts are blatantly inconsistent with the traditional values associated with military service, the Uniform Code of Military Justice (UCMJ), and the organizational structure upon which most cohesive concepts surrounding the military have evolved.[11] This provides a springboard from which to discuss female Marine cohort differentiation occupationally and socially before, during, and after this current and evolving generation.

Historically, female Marines have served alongside their male counterparts to the fullest extent allowed by law and often beyond, especially in Iraq and Afghanistan. However, they receive no recognition publicly as "real" warriors or as "real" women. They exist in a gray area, an enigma that the military and civilian communities have yet to qualitatively investigate and, much less, begin to truly understand. Who and what are these female Marines?

Attraction & Retention

The controversy over Senator John Kerry's 2006 comment, "You know, education, if you make the most of it, you study hard, you do your homework and you make an effort to be smart, you can do well. And if you don't, you get stuck in Iraq,"[12] raised important questions about the American all-voluntary military, and focused attention on whom our military attracts and retains. The senator implied that people in the military are ignorant, or dupes, or both. This research, however, asserts that, in the United States Marine Corps, the opposite is true. Cpl Alecia describes her experience living with people who think like Senator Kerry, but she believes that military service provides fulfillment that civilian life cannot:

> Where I'm from [a small town Connecticut], *nobody* joins the military. It is *so* not cool, and they're all anti-military. They're very liberal and the military is looked at as a dumping ground for losers. I was raised with my dad talking about his pride in the Marine Corps and I think it was reflected in his life. He's a good, honest man who's about more than just making money and having status in the community. I was admitted and started school at George Mason University, and I quickly decided that was *not* what I wanted to do. I majored in music, but the coursework was so unfulfilling. Even at the very beginning as a freshman, the other kids were all talking about the orchestra and director jobs they were going to have when they got out. It was all about climbing the ladder and making money and a name for themselves. It just seemed so shallow. I wanted to be more than that.

MSgt Madeline from North Dakota enlisted in 1985. She also grew up in a small town, but the expectations there were quite the reverse of Cpl Alecia's, the expectation was settling down:

> I wanted to get away from home. I wanted to see the world and experience things. Where I was raised, it was predominately white. I didn't have any contact with people from other backgrounds. My family didn't travel much, and I just wanted the opportunity to experience things. Even now, my folks will travel to see me and my younger brother who's in the Army, but I can see that they're very limited and very close-minded in their thinking about a lot of things. I didn't want to be like that. I wanted to go out and be able to say that I'd gone places and done things. It was such a

small town. I didn't want to end up marrying a farmer and having six kids because there were no options to do anything else. That's what was expected. That was the norm back there. You married your high school sweetheart. You settled down and had a bunch of kids. You never went anywhere or did anything. I didn't want to be like that. I wanted to be something bigger than that.

While the decision of committing to military service can be difficult for anyone, it is especially complicated for women. The Marine Corps environment is, quite obviously, male-dominated and is often perceived by the unacquainted as openly hostile toward women. Further, the presumption is that women in this environment are more "masculine" than "feminine." These presumptions are simply not true.

Compromising Expectations

In the mid-1980s, Christine Williams found that women who join the military are rarely motivated by a desire to defy traditional sex roles.[13] She further clarifies this statement by adding that the first-term female Marines she surveyed hoped to balance family with career. One may argue, however, that they *do*, in fact, desire to defy traditional sex roles by their very choice of military service, especially in choosing the most "masculine" service, the Marine Corps. One must remember, also, that Williams' sample included a large number of recruits who had not yet entered the gendered workforce of the Marine Corps. This research asserts that the defiance of traditional gender roles by female Marines may be sublime or overt, but is certainly present in their decision to be U.S. Marines.

To frame the discussion, sociologist Lynn Carr studied tomboys, writing that participants in her study recalled "being instructed or shamed to conform to traditional femininity – in dress, appearance, posture, manner, interests, and dating."[14] Carr further determined that the tomboys had awareness in their youth of the disadvantages of femininity (e.g., perceived weakness, physical inactivity, etc.) and the privileges of masculinity (e.g., power, freedom, etc.). This sentiency prompted them to reject the socially constructed gender identity expected of them in pursuit of that which they desired as individuals. The gender identity of the tomboys developed from their agency and resistance to the norm rather than from submission to their biology and the associated socially constructed gender expectations.

All generations of female Marines interviewed displayed similar tendencies in their choices to reject expectations or at least actively compromise them. 1stSgt Camille from California explains that she enlisted in 1989 to secure her independence:

> I wanted independence. I grew up in a real strict Hispanic household, and our father wouldn't let us girls go unless we were married. So, I went behind his back and went to a recruiter. College was really hard for me because I had to work two jobs and I really couldn't even stay awake in class. So, I joined the Marine Corps to secure my independence and to help me get through college…My father wasn't keen on me joining the Marine Corps. He thought I was crazy. I didn't look at the other services. I only considered the Marines because it was the most challenging, and that's what I was looking for. I was looking for a challenge in my life. I found it, and I've loved it. [Laughs]

LCpl Sadie from Ohio went to Boot Camp in July of 2006. She describes having the desire from a young age to step beyond traditional gender roles:

> When I was little, I always wanted to blow stuff up. My brother and my sister are in the Navy. I'm just the first Marine. I knew it was the hardest. I was a bad ass. I played football on boys' teams growing up. So, I went to the Marine recruiter and I said that I wanted to blow stuff up. He said that, because I was a girl, I couldn't. I said, "Well, I'll go next door to the Army 'cause they'll let me blow stuff up." He said, "Wait! We'll find something for you," and he did…Other girls want to be safe and stay at home. I wanted to go to Iraq. I wanted to be out there, experiencing what other people are only willing to watch on TV. If I die there, at least I will die doing something that I care about. I'll die knowing that my fellow Marines will miss me and respect what I've done. That would mean a lot to me. I was there and it was incredible.

Captain Caroline, born in Jamaica, the daughter of an American foreign services officer and a French mother, felt adrift in the world but knew she did not want to follow her mother's model:

> From a very young age, I had an idea that I wanted to be a Marine. I wanted to live outside of these normative expectations of what it

means to live life as a "woman." I looked at adult women around me, my mom who[m] I love dearly and her friends, and their lives kinda sucked. They were bored and they were boring. They did what was expected of them. They followed their husbands around from one assignment to another. They threw the polite dinner parties. They made sure the children didn't do anything embarrassing to the family. It was a lot of work and a lot of responsibility, but they were always just in the shadows of their diplomat husbands who were the thinkers and the doers, the news makers. I knew that I never wanted to be like that. I wanted to be a doer and I wanted to receive recognition for what I did. The sense of adventure was also a huge draw. Living outside the boundaries of "normal people" was what I wanted. Not just as a woman, but also just because the Marine Corps experience was so far beyond the day-to-day lives of everyone else on the planet. How many American men can say they've done the things I do as part of my job? Not many, and I like that being a Marine sets me so far apart from this other American life reality.

Many female Marines join in the spirit of resistance, soundly rejecting the expected social trajectory of embraced femininity, its expectations, and its consequences. Others, however, join in the spirit of adventure, yet another form of resistance for young women. Still others are running *from* something negative in their lives or *to* something positive that they feel is missing.

Rejection of a Personal Hell

Some young women join the Marine Corps from what they consider a personal hell, whether financial, physical, or emotional. The Corps offers these women a sense of security and personal opportunities they believe inaccessible to them in civilian life. The decision to join is particularly dramatic given the commitment embraced, but also liberating after the lives they have endured and attempted to manage with extremely limited personal resources. One example is SSgt Amber from Arizona. A single mother of three little girls, including one with special needs, Amber was working three jobs in 1997 trying to keep their life as a family together. Unfortunately, that wasn't enough.

The State came in and said that they [the girls] were spending more time in daycare than they were with me. So, they took my girls away from me. I was just working so hard and not getting anywhere. I had been taking college courses at the community college. I was sitting in class one day waiting for the professor to arrive, and was listening to a conversation two other girls were having. One of them was saying, "Well, I don't have to work because my baby's daddy does this and that, blah, blah, blah." They were talking about all of the different ways to get out of work to stay home and do nothing. I was, like, "Oh, my God! I don't want my kids to end up like these girls." I just wanted so badly to get out of the place where I was. I didn't want every day to be this constant struggle for survival, and I had to escape this constant nagging belief that I was nothing but white trash and would never be anything else….The Marine recruiter said, "This will be the hardest thing you've ever done, but when you've completed it no one can ever, ever take it away from you." I was sold on the idea of the brotherhood. I was sold on the concept of working for what you get. I was sold on that concept of *earning* your place in the group. I wanted to belong and I wanted it to have meaning. I *needed* it, and I knew it would enable me to take care of my girls.

Rather than running *away* from their former lives, similar to Captain Caroline, many young women run *to* the Marine Corps seeking a challenge, a break from the mainstream, and the personal gratification garnered from doing something that so few Americans, much less American women, do: be a United States Marine. Instead of floundering through a civilian world in which they feel they don't quite "fit" or where they simply struggle to survive, the female Marines interviewed desired more out of life, acted on that desire, and were rarely disappointed.

Seeking Confidence & Completeness

The Marines admit that they never promise anyone a rose garden. Instead, they challenge: Do you have the heart to be one of us? Recruiting language and tactics prompting soul-searching often provides the catalyst for many young women to join. One example, Cpl Jewel, a self-described Navy brat from Virginia, enlisted to improve her self-image:

I joined the Marine Corps because I always had a poor self-image. I never felt like I was ever really part of a bigger picture. I felt like I had no talents and nothing to offer the world. When I realized this in high school, I felt compelled to join because my family has a strong military tradition. It seemed like a place I could really contribute. My grandma researched it and we've had family serve all the way back to the Civil War and even in the War of 1812. I feel like I was bred for it and being raised in a Navy family, I feel like I was raised for it.

SSgt Miranda from Cedar Rapids turned down a full scholarship to the University of Iowa to enlist:

I've always wanted to teach. I received a full ride to the University of Iowa on a music scholarship, but that's not what I wanted to do. Sometimes your heart and your talents just don't match up and you have to choose which to follow. So, that left me confused with nothing truly satisfying going on in my life…I play the euphonium, which is a baritone baby tuba. Initially, I auditioned for the Army, but the Marine recruiters heard about it and they approached me, too. My dad was a Marine, actually. I asked him why they would want me. He said that the Marines have a band and they're the best in the world. I turned down the Iowa scholarship and joined the Marines because they said, "It's not about what we're going to give you; it's about what you will give yourself. With us, you will find that inner direction, that leadership and discipline to carry you through life." I was *totally* sold… I think that's the appeal of the Marine Corps. I think that's why I joined. Being good enough to be a Marine is the ultimate challenge. I was afraid of it and I did it to prove that I could. The rest is just icing.

Similarly, Sgt Rocio from San Antonio, Texas, describes coming to the Marine Corps seeking more than a dead-end lifestyle:

When it was time to graduate from high school, I noticed my friends just doing the same old things, messing around in the neighborhood, not thinking beyond the next weekend's parties. No one was going to college because our parents couldn't afford it, so it looked like it was going to be just the same thing every day until we died. Party. Get married. Have babies. Get old. Die. I thought that there had to be, that there *should be*, more than that. So, I didn't want to be like that. I knew the Marine Corps offered the ultimate challenge. There was no other way to break that barrier into a life of meaning instead of a life just leading to death. [That] way was the easy way. I needed more. When I started talking about it, people were like, "You'll never make it." That only made me want it more, to prove that I would get away and make something of myself.

Many female Marines include the desire to serve their country as one of their reasons for joining. Reciting the oath "…to support and defend the Constitution of the United States against all enemies, foreign and domestic" instills the realization that one has not just completed a "normal" job interview. The stakes are much higher in war than peacetime. The war on terrorism further complicates occupational decisions, but that has not inhibited at all female recruitment into the Marine Corps.

A Sense of Patriotism

Like the men who join, women who join the Marines often cite patriotism as the motivating impetus. This holds true over the generations of female Marines documented in this research. Sandra, for example, describes leaving a prestigious secretarial job at a private college in Southern California to play an active role in World War II:

> I had a very nice job in 1943, but one day I was in the post office and there was a poster of a beautiful lady Marine. It said, "Be a Marine and free a man to fight." I went home and thought about that, and I decided that I could probably free a couple of them. I really wanted to do something worthwhile with my life, so I went down to Los Angeles and joined the Marine Corps. It was the 2nd of August 1943. I wanted to help my country and free as many male Marines as I could to go beat the Germans and the Japanese.

Another WWII veteran, Irma, describes enlisting a year later, having to wait to be old enough, in 1944:

> I was waiting for my 20th birthday so I could join the Marine Corps. It was just more challenging than the rest of the services, and I was drawn by their *esprit de corps*. It seemed like there was a lot of comradeship, a lot of passion, and a lot of action. There didn't seem to be a lot of resentment about women coming in. I was looking for camaraderie with men. I was never at ease with other women, but I could talk a blue streak with men in their language. I wanted to be with them in that environment, fighting that war. I couldn't imagine a more important place to be at that time.

Fast-forward sixty years and the situation sounds eerily similar but with a modern twist as Cpl Jenna from Colorado describes choosing to be part of the "solution":

Basically, the war broke out and I was seeing things on the TV and was thinking hard about going to join and help them out. It was after September 11th and then the end of the Battle of Fallujah. It was New Year's Eve, and I was at a concert down at Ohio State University. A bunch of people were talking trash about how our government sucks and does bad things and how the military sucks and just wants to kill everyone. I defended the military, especially because here we were at a concert smoking weed and drinking beer and they were over there in Iraq eating MREs and probably living in tents. I decided right then and there when those other kids were being so selfish and smug that that wasn't the America I wanted to be a part of. I wanted to be part of a solution and not part of the problem.

Similarly, Captain Maya, born in New Mexico, describes her compelling desire to serve:

As a philosophy major, we're always talking about: does one person really make a difference? And, most philosophers believe, yes. If one person walks across a lawn, doesn't stay on the sidewalk, does it hurt the lawn? No. But, if multiple people do, eventually, you'll end up with a dead lawn…Well, conversely, if one person serves, they do make a difference because if no one served, we'd all be screwed…So, I wanted to serve, to do my part, to contribute to the whole. So, I figured that I'd do it in the military. It's funny, though, that I grew up practically adjacent to Quantico, but didn't know anything about the Marine Corps. All the time, I was, literally, five minutes away. My heart drew me to what I could have seen with my eyes just across the interstate.

The stories and commitment of these women represent and echo those of many others in this research. Introspection inspired their decisions. Character inspired their actions.

Inductees as Viewed from the Leadership

Perspectives can be quite different, however, whether one is looking out from or looking in on a situation. In regards to initial training, the perspectives from enlistees and those conducting their initial training remain fairly consistent and often incredulous, both positively and negatively. From a Drill Instructor's perspective, CWO3 Leona from the state of Washington describes one of her recruits while on the Drill Field before she picked up Warrant Officer:

It's funny what drives males and females into or away from college and then into the Marine Corps. For me, it was that my parents couldn't afford college and I wanted something different to do. It was that simple. At 17, I figured four years wasn't a big commitment. I wanted to travel, but you can't do that, either, without money. I had a female recruit, though, who was a millionaire. She had, like, $8,000 a month to live on, her "allowance." She'd been kicked out of, like, five or six different colleges. But, when it was time for her to graduate recruit training, I told her I needed her flight itinerary and she said, "My daddy's going to pick me up in his plane. I can give you a tail number." I could not believe it! Later, I saw her at MOS school, and she flew her whole class to New York for a weekend and put them up in a snazzy hotel. She bought a condo at her first duty station in Southern California. I was like, "*Why* are *you* in the Marine Corps?" "Daddy was a general." I guess it was kind of a get-back-at-daddy thing or some warped continuing-the-tradition-thing. The best thing about it is that, from what I've seen, she's a damn fine Marine. [Laughs]

Series Commander, Captain Kai, an MP officer from rural Minnesota, speaks to American culture and the types of recruits who arrive at Parris Island:

As a society, we've become so diversified. Our kids are growing up very "me"-centered and feeling very entitled and we see that in the recruits that join the Marine Corps and arrive at Parris Island with this attitude of, "Okay. I'm here. Entertain me." And, that's our crop of "good" kids. They come from self-esteem building schools and sports programs where there are no "losers," believing that everyone has something "valuable" to contribute...Of all of the people in America, there is a very small demographic that is affiliated in any way with the military. Americans haven't been asked to really sacrifice or to really feel or think about our international policy in a very, very long time. I think Vietnam was an anomaly because that was the first "war" that was put in America's face via a TV screen. Since then, though, we've become more and more desensitized to watching brutality on TV and in the movies. September 11th was another anomaly because it was one incredibly horrifying real-life event. As we watch it over and over, though, it becomes less of a tragic event than just more visual stimulation played to evoke emotion. It's really sad. Of all of those crazy magnetic "Support our Troops" and "God Bless our Troops" on so many cars at the beginning of the war in Iraq, how many are now so faded that you can't even read the words. And, they're not being replaced, they're just being forgotten, just like the American and international service members that are over there still trying to help the Iraqis make a better life for themselves without a dictator. After all the initial emotional outpouring of 9/11, no one seems to care anymore. It's true for a lot of these kids that arrive on the yellow footprints. Most of them, I'm just not convinced that they've joined for the right reasons.

On an even darker note, Series Commander Captain Caroline is troubled by what she physically sees at recruit training:

Living every day with recruit training was very hard for me. Doing the initial hygiene inspection was worse than I could ever have imagined. These young women have scars from bullets and knife fights and tattoos that are gang related. Their fathers, mothers, or boyfriends beat them, or they are, like, 18-years-old and they have Caesarean scars! It's such a sad atmosphere. I felt like I was too naïve to be there. Here were young women who would, potentially, be in my platoon or company someday in the Fleet, and I felt like I had nothing in common with them. I was amazed by their bravery to make the decision to become Marines, but I was also repulsed by what I'd seen in their young female bodies. Mentally and emotionally, I wasn't sure what to do with that information.

As further support, MSgt Darla from Maryland tells of her experience as a Drill Instructor, seeing so many young women running *away* to the Marine Corps, and explains why separate recruit training for the sexes is so important:

I think that, initially, when we enter the Boot Camp that we should be trained separately. We do it the right way. And, the reason I think that is that it's the shock factor. I think they also need the shock with somebody of their own sex that they'll be comfortable with. One of the things I found, that I learned, was that some of those girls when they came to recruit training, they were running *away* to us. They could have been running from their father or their brother or another family member who molested or otherwise abused them. They were running away from being on the streets. I mean, who knows what they were running from? Some of them were seeking to feel like they belonged, and I think that, if it was a man yelling at them when they got to Boot Camp, many of them wouldn't make it because what if that happens to be that recruit who was trying to get away from that man who molested her or from the streets where she worked as a 14- or 15-year-old prostitute? Some of the stories I heard were just *unbelievable*.

Feeding the enlistment machine, GySgt Hailey describes a humorous but also disturbing situation while on recruiting duty in the late 1990s:

When I was on recruiting duty, we had an all-female poolee function at RS New York and the young women kept asking questions like, "What if I get pregnant? Does the Marine Corps pay for that?" Or, "What about marriage?" These were 16- and 17-year-old girls. They kept asking all kinds of questions like these, and the couple of female DIs that came up to speak to them and I was *[sic]* all like, "*Whoa*! If you're thinking of joining the Marine Corps to get a husband, you certainly can do that, but there are much easier ways to do it!" [Laughs] "Go to college instead!" It was really crazy to actually hear that kind of mentality.

As a Recruiting Station Operations Officer, Captain Sophia also sights in on why young American women choose to join the Marine Corps:

I think so many of the girls that we enlist are looking to grow into something special, something that they don't believe the civilian community can give to them. Some of them are looking for a completely different start because their situation hasn't been what they'd expected or hoped for. They're looking for the *esprit*, the bond, and the confidence that they find in the Marine Corps. We have all of that, but they must earn their place among our ranks. It's not an easy process by any means, but for the right kind of person, it's definitely worth it.

American females have the opportunity to run *away from* or to run *to* various options available to them, including military service in the Marine Corps. Motivations include financial, rebellion, social/sexual access to males, and patriotism. All, however, exemplify a conscious choice to be different from females in the American mainstream. These data suggest overwhelmingly that female Marines *do* actively defy and/or manage their biological sex while also selectively rejecting the gender roles defined for them by society. In doing so, many endeavor to create a synthesis of male and female social roles, of power and submission, supported in very obvious ways by this research. Expecting to be a wife and mother is but one aspect, as Williams found and GySgt Hailey described. However, those roles emerge as ancillary to the more versatile *self* that a young woman fashions as a U.S. Marine. A liberating concept such as this, however, can also be viewed as confining, as it is likely that she will find herself more encumbered as she is expected to perform successfully *all* of the various masculine and feminine roles she has attained.

From their own testimonials and observations, one can begin to understand the motivations of young female Americans who choose to join the United States Marine Corps, an acknowledged male warrior preserve and an alleged den of misogynist angst. The findings of this research reject Moskos' and Meštrović's concerns for the future of a postmodern and postemotional military devoid of emotional and professional commitment, although the form it takes may present differently. Female Marines continue to enter the Marine Corps for a variety of reasons, including patriotism. They bring with them personality, skills, energy, sometimes scars, and often a keen awareness of the difference between a life without meaning and a life with purpose. Instead of remaining on a preordained, socially normative path, and perhaps without even fully understanding their motivations, they embrace Marine Corps core values, the physical, mental, and emotional rigors of the military experience, and 240 years of proud tradition.

Whatever their intentions, how they enact these motives and values after initial training, however, is another matter. To earn the title, "Marine" and experience the Marine Corps lifestyle a young woman must first prove herself worthy via Recruit Training, affectionately called "Boot Camp" or Officer Candidate School (OCS) and The Basic School (TBS). These transformative processes remain unlike any other on earth, and the journey begins on the yellow footprints of Parris Island or in Quantico, Virginia.

Endnotes

1   Charles C. Moskos, "Toward a Postmodern Military: The United States," in *The Postmodern Military: Armed Forces after the Cold War*, eds., Moskos, Charles C., John Allen Williams, and David R. Segal (New York, NY: Oxford University Press, 2000): 19.

2   Thomas E. Ricks, *Making the Corps* (New York, NY: Touchstone, 1997): 19.

3   Mitchell, *Women in the Military*, 33.

4   Culver, "Women in the Service," 65 and Shields "Sex Roles in the Military," 99-114.

5   Sandra L. Myers, *Westering Women and the Frontier Experience 1800-1915* (Albuquerque: University of New Mexico Press, 1982), 98.

6   Moskos, "Postmodern Military," 27.

7   Stjepan G. Meštrović, *Postemotional Society* (London: Sage Publications, 1997), xi.

8   Ibid., 43.

9   Ibid., 38.

10 Mitchell, *Women in the Military,* 331.

11 Wakin, "The Ethics of Leadership II," 201.

12 Jim VandeHei and Chris Cillizza, "Bush Calls Kerry Remarks Insulting to U.S. Troops, *The Washington Post* on November 1, 2006, A Section, Page A08,   http://www.washingtonpost.com/wp-dyn/content/article/2006/10/31/AR2006103100649.html.

13 Williams, *Gender Differences*, 73.

14 C. Lynn Carr. "Tomboy Resistance and Conformity: Agency in Social Psychological Gender Theory," *Gender and Society* 12 (1988): 543.

Chapter Three

Initial Training: Is Separate Inherently Unequal?

The military is a unique institution, and women comprise less than 7% of the total force of the Marine Corps. Marines represent a diverse and elite microcosm of American society, required to meet stringent physical, mental, and occupational standards. Since 1775, the Marine Corps has evolved as have the expectations of its female members. For at least four decades since World War II, female Marines were expected to conduct themselves as ladies and not as military leaders. Instead, their training included required make-up classes and application, mandated leg shaving, and formal instruction in behavior at tea parties. Today, the Marine Corps is the service with the most physically and mentally demanding expectations of its females, inching ever closer to equal physical fitness training and testing standards and the inclusion of females into combat MOSs in 2016.

Meeting continued failure to full integration, however, are female Marine officers' completion of the Marine Corps' 13-week combat Infantry Officer Course (IOC) at Quantico, Virginia, and implementation of dead-hang pull-ups for the female Marine Physical Fitness Test (PFT) to replace the flexed arm hang.[1] These failures should come as no surprise although, somehow, they do to the organization as well as the media. Authors such as Colette Dowling argue that through intensive training females can achieve equal physicality to males.[2] This reflects the spirit in which Drill Instructors at Marine Corps Recruit Depot (MCRD) Parris Island, South Carolina, continue to train female recruits for physical success although the pull-up mandate has been delayed until the end of 2015.[3] The data offered here support female Marines' on-going struggles with achieving upper body strength and endurance equal to that of male Marines. Their physiology limits their capacity to perform some physical tasks much more easily accomplished by the male body. Their physical limitations and the associated perception of inferiority remains a source of frustration and embarrassment for female Marines and profoundly impacts their leadership potential and career success.[4]

Leveling the Playing Field

Female Marine 2nd Lieutenant Sage Santangelo, one of the females to date to attempt and not complete the IOC, alleges that the Marine Corps sets up females for failure by not training to the same standards as male Marines. Her assertion is that segregation by sex and separate standards in initial training do not allow nor encourage female achievement of the required level of physical fitness and performance. One immediate victory for equity emergent from Santangelo's public proclamation is that females are now allowed to retake the IOC as males historically have been allowed to do. Of note, as well, is that the Commandant of the Marine Corps, General James Amos, "offered Sage Santangelo a posting *[sic]* in Afghanistan while she waits for a flight school opening." The General's footnote to the article raises questions about Santangelo's career aspirations and reinforces the quandary created by allegations of favoritism bestowed on females in the military as forms of seduction or appeasement.[5] On a positive note, enlisted female Marines have achieved success consistently in the less rigorous infantry training program at Camp Geiger, North Carolina. Although satisfactorily completing the training, they cannot assume the related combat specialty of infantryman until the combat exclusion ban is lifted.[6]

Historically, the Marine Corps has exhibited good faith responses to civilian demands for sex/gender equity in physical training and testing. Female physical fitness testing, completely separate from the males, was mandated in 1968. The run portion of the female Physical Fitness Test (PFT) increased to three miles via Change 2 to MCO 6100.3J of 3 November 1997, matching the required distance of the Marine Corps males, albeit with a variation on required time; females are allowed one minute more per mile than the males.[7] Also in 1997, female recruit training was extended by a month to include additional weapons training and "The Crucible."[8] Although not allowed in combat MOSs, they must also successfully complete the pass/fail

Combat Fitness Test (CFT) with the same tasks as males but a different grading scale.[9] In the past decade, the Marine Corps implemented the gender-neutral "One Mind, Any Weapon" martial arts program (MCMAP) to enhance physical fitness and occupational competency. The belts awarded by completion level, indicators of commitment and proficiency, may be worn in lieu of the prescribed uniform belt.

Implemented over the past few decades, these and numerous other attempts to eliminate physical, training, and occupational disparities and the resulting friction between males and females have met with questionable success. Lieutenant Santangelo argues that the current system fails female Marines striving for leadership and thereby occupational success with their male peers. Both biology and social constructs render the transition from civilian to U.S. Marine complex and challenging for many young female Marines. The qualitative data presented herein provide historical context and offer mixed beliefs and opinions about the legitimacy and functionality of separate initial training for male and female Marines.

Vicarious vs. Actual Status Attainment

American society's history includes a sex-based division of labor. Since the 1970s, however, its young females actively have pursued voluntary military service. One might assume then that American women now possess every opportunity to achieve the same level of occupational success and social freedom as males. The reality, however, is that biological differences, gender preferences, and persistent social expectations remain. The reasons are layered within humans' evolutionary development, social constructs, and Free Will.

Evolutionary psychologist David Buss explains why females attempt to secure mates with superior resources: "Women desire men who command a high position in society because social status is a universal cue to the control of resources."[10] Since a mate with superior resources likely ensures a higher social status for her and their progeny, a female generally will do better to secure a mate with resources rather than attempting to secure them on her own, especially while also caring for children during which time her resource-attainment options may be limited. This is likely the subconscious motivation of the young females GySgt Hailey described and reiterated by other women in this project.

From a sociological standpoint, a complimentary argument has been made. Veblen identified commodification of women who reflect the prowess and social value of the men to whom they are mated.[11] Thus, women devalue their own potential to contribute equitably in relationships with their mates when they assume submissive and objectified statuses, relinquishing their sexual freedom to adopt maternal roles, biologically and socially.[12] For female Marines like SSgt Amber in the previous chapter, the two endeavors of competing for resources and raising children do not correspond positively socially or financially. She elected, however, to pursue a professional career as a single parent and struggles to reconcile competing demands. Rather than attempting to claim social status vicariously through men, still quite common in our "liberated" society,[13] female Marines like SSgt Amber adopt a lifestyle that offers them perhaps the most unrestricted opportunity in the U.S. This is the opportunity to be physically, professionally, and occupationally equivalent to their male counterparts.[14] This lifestyle, however, is also rife with vicariously attached "camp followers" who undermine the achievements of higher-performing female Marines.[15]

As previously asserted, female Marines attempt to break the barriers to fulfill their needs. SSgt Manda describes a demoralizing incident that occurred a few years before her decision to enlist in the Marine Corps, but relishes the validation she experienced because of her decision:

> When I was in college at Texas A&M, my roommate was dating a guy in the Corps of Cadets, a Marine option, so he was on the fast track to a successful Marine Corps career…She said, out of nowhere, "You know, you could only ever be an enlisted man's wife." I was stunned, and my first thought was, "You self-righteous bitch." [Laughs] Using nasty words like that, I guess I probably deserved her flip remark, but I do know I didn't deserve her judgment of me at all…She was separating me from the military "upper class" she knew she was entering when she married that doofus. A few years later, I saw her at Camp Lejeune from a distance loading a couple of kids into a car there in Paradise Point, the officers' housing, and a part of me really wanted to hate her, but I couldn't because I was on my way home from PT after running the Obstacle Course a few times with my company. I knew *I* was the real deal and she was just living the life vicariously. I had

done even better than being an enlisted man's *or* an officer's wife; I had earned the title "Marine" for myself!

Wearing her own earned rank, the Marine Corps is an environment where a woman can be her own person without the initial association of being a particular male's wife, daughter, sister, etc. Historically an all-male preserve, the alleged patriarchal social structure based upon sex and the relationship of the sexes to one another has been broken down in the Marine Corps.[16] Ideally, equivalency should be achieved via a rank structure that determines one's status. The better one performs, whether male or female, the higher status one can achieve.[17] Warfighting, however, is a complicated and demanding occupation. Sex/gender further complicates this.

Warrior as Gender-Neutral

Martin van Creveld states, "Of all activities war is by far the most nasty and the most dangerous. It is also physically the most demanding, which means that, despite the claims of some feminists to the contrary, in no other activity are women as much at a disadvantage in relation to men."[18] It is delinquent that he excludes childbirth as one of the "most nasty and the most dangerous," and one from which men are wholly excluded. Apparently, the element of *choice* in killing one's enemy differs so largely from the *imperative* of a woman carrying a man's child within her to propagate the next generation that renders the latter irrelevant and, thus, warfare infinitely more glorious.[19] Captain Caroline speaks directly to assertions such as van Creveld's. She adamantly professes simply that, because men monopolized the endeavor for eons, they are not necessarily "superior" in the warfighting environment:

> Stress and physical differences between men and women are a concern because that's ultimately what combat relies on: the physical, mental, and emotional prowess of combatants. When people's lives depend on a Marine's conditioning, things like physical strength cannot be discounted as "unimportant because her job is this or that." Today's battlefield, unlike that which was imagined by Star Wars thinkers and politicians, does not automate everything. It's physical and it's dirty and it's very, very personal every day. I've *lived* it. Some women are very well suited for this, just as some men are. Some will fail just like some men will fail,

and there have *always* been males who have failed. Why do the historians overlook that and presume that "warriors" are these supermen that *always* overcome? If that was the case, no man would ever have died in battle, and guess what? A *lot* of men have died in battle! It's just a fallacy like so many other things we've been taught over centuries about male-female relations to maintain the status quo.

CWO4 Renae also questions alleged male superiority in combat, attributing "luck" to many events in war[20]:

Women deal with combat just as well as the men do. High-stress situations require nothing more than absolute control of one's emotions, attitude, and behavior. A lot of what happens in combat is luck. So many historians want to overlook that. Why, of two Marines equally prepared for the engagement, did one die and the other not? Luck. One sniper or bomber got lucky. It's just that brutal and that true. We can train and train and train, but sometimes Fate just has her way with us, and we as humans hate that…A Marine is going to take care of his squad-mates and his platoon-mates, and his company-mates and his battalion-mates, pretty much in that order because that's the hierarchy of who he lives with every day. Contrary to all the media hype, the so-called "male bonding" does not always involve hookers and shit like that. It's much more subtle than that. These guys live together like in no other environment. They are, literally, intimate friends who know each other like few "normal" humans will ever know other humans. That doesn't mean that they're *sexual* friends; they're brothers and best friends. They're "you cover my ass, and I'll cover yours" friends in a world where their asses do, *in reality*, require covering! The women, the female Marines, who are in their kinship network are also part of the male Marine's life, and are not viewed any differently than another beloved and trusted male. The keys to a successful relationship are full integration as an equal partner and the group's confidence in a Marine's integrity and occupational competency.

Beginning the Transition to U.S. Marine

As these two Marines assert, entering this warrior class requires an initial test of mind, body, and spirit. Female Marines begin a journey seeking self-fulfillment and status attainment on the yellow footprints at MCRD Parris Island, South Carolina, for enlisted Marine recruits or Quantico, Virginia, for officer candidates. Captain Joy, a Series Commander at Parris Island, identifies a key difference between female Boot Camp and OCS and why that difference exists:

> Something I have learned being at Parris Island is that there will always be that bottom 10%, that group that will finish last and sometimes under questionable circumstances. Because of the opportunity to DOR [Drop on Request at OCS], I believe that the vast majority of women who graduated actually deserved to graduate. There were no waivers. There were no quotas. The situation was, if you don't wanna be here, leave. That's something that's especially difficult to deal with at recruit training. You have recruits that obviously don't want to be here. They know they're in over their heads, but we still try to motivate them and give them the skills they need to succeed. It's really a blow to the ego, too. Why don't you want to be in my Marine Corps? Why don't you want to be here? I have struggled with the difference in the two groups. What I understand is that, with the officer corps, the women there at OCS have already gone through ROTC, they've been to Bulldog, they've already been enlisted if they're ECP or MECEP, or they've just had the opportunity of being in college to get an idea of what they're getting themselves into. If you're a female and you're trying to sell yourself to an OSO [Officer Selection Officer], you better be damn good. They'll tell a female, flat-out, "No, thanks." In the officer ranks, wanting to be there and wanting to lead is just too critical. The saying goes that a bad officer is gonna get a lot of Marines killed, but there's always a job for a bad PFC. It makes sense. An enlisted person can contribute their four years doing something for the Marine Corps as long as they're not total delinquents and they can go on their way. A bad officer will have far-reaching negative effects, though, if the honor, courage, commitment, integrity, and all those other things critical to Marine Corps leadership aren't there.

In contrast to having an ideal and a plan, Captain Dawn, a Logistics Officer, describes arriving at Quantico clueless about her status and the expectations of a Marine officer:

> I didn't really know there was a difference between officer and enlisted. I just knew that some recruiter, some OSO guy, was, like, "Hey, we're gonna send you to OCS," but that really didn't mean anything to me. I was just, like, "Oh, I'm going to Boot Camp,

kind of." And, when I got there, they were, like, "You know, we're gonna be officers," but I was, like, "Okay. I'm just gonna be a Marine." Some girl clued me in. She said, "You have to salute the shiny stuff. The black stuff is what yells at you now, and then works for you later. We're gonna be the shiny stuff." And, I was, like, "Okay," and, I guess, it started to click. It wasn't easy. I'm not saying it was easy or anything, but I did think it was gonna be a little bit worse. About half-way through, I was, like, "Okay, I can do this. I think I can make this." My female peers were the most challenging, learning to get along with all those other females. That was really hard.

Captain Aurora, a Military Police Officer, agrees with Dawn that OCS as a training program was extremely challenging in all its aspects:

[OCS] was hard, but it was what I wanted. Honestly, the hardest thing was living with so many other women. The cattiness, the possessiveness was almost too much to stand. The "politeness" went away after about a week. It was everyone living in their own small space, trying to figure it all out. There were those who were weak and couldn't decide if they wanted to be there, and the strong ones ended up trying to carry them, at least until they washed out on their own. There was the bitchiness, the "Leave me alone!" But, that's not an environment to be alone in. It just doesn't work. Even the training in ROTC, I was one of only two females amongst about 20 males and the attitude there is just to get the job done and it's over with. When you're with females, though, everyone has an opinion and they all want to share their opinion before a decision was made. So, everything became a drama with one person getting her feelings hurt and another female trying to build an empire, while yet another female was hanging back and being a problem because she didn't agree with any of it. It was insane and I hated it! At the same time, though, I met a couple of life-long friends at OCS. We just developed a personal and professional bond, mostly because we shared the opinion that most of the women there should have stayed home.

Captain Caroline, a Communications Officer, expresses the fears she felt coming to the Marine Corps from the French school system and traveling around the world:

When I got to Quantico in summer of 2001, I was ready physically and mentally. I was very strong in the upper body for a female, which really helped. Emotionally, I did fine, but I kept thinking that I wasn't worthy to be there. I was thinking, what made me think that I was good enough to lead Marines? I was actually over-awed by the experience. It seemed to me that everyone else was so much better prepared or knew more than me, or that they were natural leaders and it came easy to them while I was working my ass off to make it all work. During the endurance course I failed by about 30 seconds or so. I was just having a bad day and after that, I cried for the next three nights just overwhelmed by the experience and feeling like I had made a choice that I couldn't live up to. Finally, all the other reasons fell away and I began to realize that I was there to try to make myself a better person and it became so much easier. Instead of this ideal that I was trying to live up to, I started to focus on making myself better, personally and profes-sionally. That was my key, and that's been my key ever since.

Becoming an enlisted Marine presents similar challenges as those described by the officers. LCpl Patrice, an Engineer Equip-ment Mechanic, describes her experience:

The first two weeks, from the moment I stepped onto Parris Island, were terrifying. I had no idea what to do. They were yelling at me to do stuff, and I couldn't remember what they said the second it was out of their mouths! I had never had so many people yell at me about nothing. I've never been the type of person to "take it" from anybody. I always stand up for myself and others who can't, so it was hard for me. After the first two weeks, I realized that this was their [the Drills Instructors'] job, to break us down and build us back up. Yeah, they're scary! It's their job and I have my job here, which is to learn from them.

Conversely, SSgt Tomasa' background of emotional and physi-cal abuse by her mother prepared her for the mental challenges of Boot Camp, and she took the experience in stride:

The day before I left for Parris Island, one of the neighbors said, "We'll see you back in a couple of weeks." I was, like, "No way." I got on the bus and didn't look back. The recruiters showed me a video, and I knew that they couldn't hit me and that they'd feed me, so it was going to be okay. This was August 1994. I had to gain two pounds to ship. My mom was such a hard ass. She was the kind of person that you would be nervous waiting to see what kind of mood she would be in when she got home. She would be physically abusive at times, so the yelling didn't bother me in Boot Camp. If anything, I got in trouble for laughing at the other girls. I was IPT'ed all the time. I stayed up there [on the quarter-deck]. The Drill Instructors would ask, "WHO OWES ME?" I didn't even have to think about it; I knew there was something I had done, so I just went up there. I was in great shape. [Laughs]

Via the various paths available to them, these female Marines and the others interviewed cast away their civilian lifestyles and entered the regimentation of military life. Their expectations of themselves coalesced as they began to comprehend the expectations of the organization and their peers on them. Always challenging, the transition is easier for some than for others. One of the initial and most basic challenges for all is body management.

Body Management: "You Better Square Away that Nasty Body!"

Over the years, the Marine Corps' preference for its women to be ladies evolved to a preference for them to be competent warriors. CWO4 Renae describes aspects of Boot Camp in 1981 that no longer exist on the training schedule: "We had etiquette class back then…we were taught how to properly get out of a car! We were taught how to sit when having a conversation with someone. Now, females are taught how to kill the enemy and how to keep our Marines alive. Thank God!"

In keeping with the tradition of Marines as impeccable citizen-soldiers, personal appearance ranks equally with prowess as a Marine's priority. Marine Corps Order P1020.34G W/CH 1-4 of 31 Mar 03, Marine Corps Uniform Regulations, states: "Marines are known not just for their battlefield prowess, but for their unparalleled standards of professionalism and uncompromising personal conduct and appearance. It is a Marine's duty and personal obligation to maintain a professional and neat appearance. Any activity, which detracts from the dignified appearance of Marines, is unacceptable." In addition to maintaining an impressive physical appearance, a female's biology must be mastered to ensure that it does not become an embarrassment, a liability for her personally and professionally, or a point of contention between the sexes.[21] In restraining her female-ness, a female Marine becomes more like the males while also reducing the friction her sexuality creates by its presence.

Managing Menstruation

Often cited historically as one of many biological reasons why females should not be allowed in the military, in general, and in combat, specifically, menstruation poses unique challenges to a female's body management in the physically demanding military environment in garrison and in the field.[22] Menstruation for some women ceases during periods of high stress or intense physical activity, the female body's appropriate reaction to mental or physical stress that creates an environment unsuitable to pregnancy and childbirth.[23] Now, in addition to the human body deciding when procreation may or may not be sensible, in 2007, the FDA approved Lybrel™ for use in the U.S. A low-dose hormone form of birth control, Lybrel™ eliminates regular monthly periods, although unscheduled bleeding or spotting is likely, which could potentially be more awkward than the "normal" period one expects. In a society always looking for the "next big thing" in body management, however, pharmaceutical options like Lybrel™ will likely advance in efficiency as well as popularity in controlling unwanted bodily functions. Interestingly, none of the female Marines interviewed considered this issue as debilitating to or even inhibiting their success as Marines. Many asserted that all Marines in the field are just nasty, regardless of sex.

The Complication of Hair

The head shaving of male military recruits receives attention from a number of sources. For example, anthropologist James Brain states, "…in the triumph of authoritarian–the army–the recruit is always given a ritual and very often humiliating short haircut to indicate his submission to control by those *in loco patris*."[24] In contrast, scant research exists that discusses the similar humiliation female recruits endure. Evolutionary psychologists agree that a woman's hair is closely tied biologically to her sexual appeal and mating success[25] while anthropologists and sociologists agree that hair is closely tied with cultural constructions of femininity. In 1987, for example, Anthony Synnott stated, "…although hair grows all over the body, in terms of body symbolism there are only three zones of social significance: head hair (the scalp); facial hair (beards, moustaches, eyebrows, eyelashes, sideburns); and body hair (chest hair, arm-pit or axillary hair, leg, arm, back, and pubic hair). Each of these zones has both gender and ideological significance."[26] Significant in mating, Buss asserts that head hair is one of a handful of cross-culturally significant indicators of a female's physical beauty. To remove a woman's hair, then, can be equated with eliminating her potential for successful mating. The same is not as true for men because women prefer other indicators of mate-ability such as physical size, strength, and evidence of resource attainment.[27]

For male Marines, the head hair is completely removed upon arrival at Boot Camp or OCS, and they must remain clean-shaven throughout their active service; their head as well as facial hair are strictly controlled by Marine Corps Order. Also mandated by Marine Corps Order, females' head hair is not removed upon arrival at recruit training or OCS, but it can be a significant problem that must be managed.[28] Prior enlisted Marine, CWO3 Phoebe, describes the importance of hair in the mid-80s:

> My hair was the big thing and I hung in there as long as I could, but I finally succumbed and ended up with the whack job Opha Mae and her sistas [local beauticians] give you at Parris Island. Talk about your beauty school drop-outs! [Laughs] Oh, well. It was my decision, and it did grow back and by then I was in a better position to properly take care of it. It may sound silly, but it was traumatic for me as an 18-year-old girl. I had great hair, and it was a symbol of my girly-ness. It was the 80s and big hair was a *huge* part of the popular culture.

In 1988, per 1stSgt Halona, the situation remained the same:

> There was one girl who, every night, braided my hair for me. I didn't know how to braid my hair! So, she gave up her free time to braid my hair for the next day. We weren't allowed time to wash our hair in the showers. That would have been a frivolous waste of valuable Marine Corps time, so we washed our hair in the deep sink on free time, that one hour that we got every night. There were really one or two that would do it [braid], and there would be lines at their footlockers. As paybacks, another girl might iron for her or polish her boots or whatever. It was an assembly line of sisters, each doing different tasks to help each other.

Sgt Ruby describes the situation almost 15 years later in 2001:

> In Boot Camp, you had, like, two seconds to get ready in the morn-

ing and, if it was that time of the month, you figured out real quick that you needed to get up about twenty minutes before reveille to get yourself together for the morning and totally plan for the day, or you were screwed. Same thing with hair. Any of the girls that decided to keep their hair had to stay about ten steps ahead of the Drill Instructors. At first, the recruit thought she'd just stay up after lights out and do her hair. Un-uh! The DIs were all over that because lights out meant you were in your rack sleeping. So, then they tried getting up in the middle of the night and doing their hair but the fire watches busted them because it was *their* asses for letting you screw around after lights out. After a couple of weeks, it seemed like getting up before reveille, quietly doing what you had to do, and then quietly going back to your rack seemed to work best. But, you had to decide which was more important, your sleep or you hair. A lot of girls chose their hair, which I thought was funny, but it was their choice to make. It's challenging because you have to totally manage your own limited time because once those lights come on, you're on Marine Corps training time and the DIs own you.

Whether considered a symbol of girly-girlishness or of good health for breeding, numerous female Marines alleged that a female's hair is significant, biologically and socially, much more so than for men.[29] Hair maintenance embodies but one challenge to female Marines who constantly must prioritize time commitments and remain within military regulations.

Cosmetics Use a.k.a. "Image Development"

Throughout the recent history of females in the Marine Corps, the requirement to wear make-up has vacillated. For example, during the Free a Man to Fight generation, make-up was not required in uniform; this likely relates to societal practices of that era. Darcy, a World War II veteran, recollects, "We wore make-up if we wanted to, but it certainly wasn't required. Not many young ladies did back then, really." Sandra, also from that generation, remembers make-up being optional. In the 1940s, female Marines often wore Elizabeth Arden lipstick named Victory Red as well as Montezuma Red, which was created specifically for the lady Marines to match their government issued red scarf, red arm chevrons, and red cord on the service cover.[30]

In contrast, thirty years later during the AVF generation, in an era that female Marines entered MOSs as occupational peers of the males, Marine Corps Order required that females wear make-up, eye shadow and lipstick at a minimum. Labeled "Image Development," in the opinion of numerous female Marines make-up classes in Boot Camp created more monsters than ladies. For example, SSgt Lucy, an African American, describes her Boot Camp "Image Development" experience:

> By the time I got there [Parris Island in 1981], they had cut out the etiquette classes, but they still had the make-up classes. They had local women who came in and showed us how to put on make-up, but the problem with that was there were no African American women to show us and the make-up that they had was for white women. So, the Hispanic women and the black women that were there, we basically had to wear that make-up. We had the blue eye shadow. We had the red lipstick. Of course, scarlet red was the only color we could wear at the time. The foundation, the blush was for white women. So, the black women [recruits] there looked pretty silly.

Prior enlisted Captain Taylor expresses gratitude that she missed "Image Development" in 1987:

> I was on fire watch that day. I was thankful, too, 'cause the girls came back looking like streetwalkers. It was awful. I thought, who the hell's teaching these girls? Avon is *not* here. Avon *should* be here teaching these women. The "instructors," if I dare call them that, were Mayberry rejects, and had big, poofy hair and bright blue eyeshadow. They looked like Tammy Faye Baker. It was terrible, but I had to laugh, too.

These and numerous other stories about "Image Development" and the previous make-up requirements make today's generation of female Marines, incredulous, laugh out loud. Marine Corps Order regarding uniform regulations now only state, "Cosmetics, if worn, will be applied conservatively and will complement the individual's complexion tone. Exaggerated or faddish cosmetic styles are inappropriate with the uniform and will not be worn."[31]

Uniform Changes[32]: From Culottes to Kevlar

Over the past 70 years, uniform changes in the Marine Corps for females have ranged from culottes[33] to Kevlar®. As early as

1977, the Marine Corps leadership began to recognize that if the females were to be active participants in Marine Corps training with the males, adjustments would have to be made to their uniforms. One example is from Twenty-nine Palms, California:

> Integrated battalions and companies such as this one gave rise to some interesting adjustments, notably in the area of physical training. In this instance, the battalion organized a competitive seven-mile conditioning hike. The course included a climb over hills behind the main camp, but because the WMs did not have adequate boots for the cross-country portion, a seven-mile road march was planned for them to be lead by Captain Ables. The battalion commander had arranged to take her company himself. The women's platoons from each company were combined to form a single WM [Woman Marine] unit and scheduled to hike on the day before Captain Ables' Headquarters Company. Having finished her portion of training, Captain Ables was challenged by her husband, Major Charles K. Ables, to lead her own company the next day. She admitted that it was a struggle to run-walk to keep from straggling. It happened that she was not only not the last to complete the course, but she helped to push a Marine over the finish line, and Headquarters Company won the competition. Afterwards, it was decided that future company hikes would be conducted with men and women participating together, maintaining unit integrity.[34]

In addition to the issues of separate training and uniforms, this quote also identifies: 1) the challenge of a male Marine to his female Marine spouse who is one rank lower than he is, 2) the struggle of the woman to keep up on the hike, 3) her apparent pride in not being last, but also assisting another Marine complete the hike, and 4) the fact that unit integrity is compromised when training activities are separated and disparities in functional uniforms exist. At this point, the Marine Corps began to seriously consider the implication of separate standards as well as separate uniforms for the two sexes. Similar to the situation above, Major Lauren describes her Boot Camp experience in 1977, including the lack of appropriate footwear:

> The only thing I really remember about me in Boot Camp is that I couldn't iron a uniform to save my life. That's when we had the blue uniforms. When we were in Boot Camp, we drilled. That hasn't changed much except that we weren't allowed to have ri-

fles. The only rifle we saw was on a TV screen for instructional purposes, kind of a FYI. We never touched a rifle; we weren't even allowed to. Our PFT was a mile-and-a-half run and we walked out to Elliot's Beach [the "field"]. We actually *walked* out; it wasn't a "hike." We had oxfords on, leather. Basically, it's the same shoe we have now, but real leather. That's what we lived in then. It was tennis shoes or oxfords. We weren't allowed to wear boots…We did PT, which was in the culottes, not shorts…For the SDI's inspection, I remember being in front of my Senior Drill Instructor and she asked me if I'd turned my iron on. At the time, we didn't have Alphas. We still had the green pin-stripe uniforms for summer.

Starting in the early 1980s, to keep pace with the new military training requirements for females, uniform changes also came fast and furiously. Former SSgt Lucy remembers the transition in 1981 to weapons handling and utility uniforms for female recruits:

We were the first platoon at the time to have training with the weapons. They told us we were the test platoon because women previously had not trained with the M-16 [rifle]. We were also the first platoon of women to wear camouflage utilities. Before that, they wore khaki-looking outfits like what the Army wears. Remember the show *Gomer Pyle*? Well, the women wore khaki shirts and khaki skirts like that. That was considered their utility uniform. I honestly don't know if they were allowed to wear pants or not.

Captain Olivia describes what she was instructed to bring to recruit training in the fall of 1989:

The packing list for Boot Camp was insane. The pink pamphlet said that if we smoked to bring lighters and cigarettes. You gotta be kidding me!! [Laughs] Drill Instructors were really the ones who coined the phrase, "If you're smoking, you better damn well be on fire!" I was instructed to bring things like a girdle and a bathrobe, pajamas; things that weren't needed anymore. They apparently had been in the past, but not anymore, and that's all I had to go on. My clueless male recruiter didn't know what *female* Boot Camp was like. I didn't even know anyone who wore a girdle much less where to buy one!

Other significant changes to the female Marine uniform included the addition of maternity service uniforms and then later, maternity utility uniforms in the 1990s. During the Free a Man to Fight generation, pregnant females were discharged, but in July 1975 then-WMs could request to be retained on active duty, and could wear civilian clothes when the uniform no longer looked appropriate. The seemingly unlikely prospect of a regulation maternity outfit was under study by the military services and later approved.[35] In 1996, female Drill Instructors removed the scarlet shoulder cord, worn since 1983, now authorized to wear the campaign cover, "Smokey," like the male DIs. In 2007, per CMC decision and disseminated via MARADMIN 504/07, female recruits would now be issued Dress Blues upon graduation from Boot Camp.

In terms of body management and physical presentation of the organization originating in initial Marine Corps training, equitable if not identical uniforms facilitate kinship and equivalency. Although still not physically "the same" as their male counterparts, their uniforms and the expectations that accompany them continue to bring females closer to a status as equivalent members of the Marine Corps.

Weeding Out the Weak & Unworthy

The Marine Corps possesses the reputation as the toughest fighting force in the world.[36] It only follows that the females within it must be the very best, too. Even during World War II when women's standards were a world apart from men's, there was a system in place designed to eliminate the females who could not, or would not, meet acceptable standards. Sandra remembers Boot Camp at Camp Lejeune in 1943, stating, "Several women who insisted on smoking after lights out or other transgressions ended up disappearing. I don't know if they or the Marine Corps decided they were unsuitable, but they were sent back to where they had come from." The key word here is "unsuitable," and removing the "unsuitable" from its ranks has always been and remains a priority of the Marine Corps today. For example, 1stSgt Camille describes her Boot Camp experience in the spring of 1990:

> There were women who shouldn't have been there. They were there for the wrong reasons. I was a little older than most. I was 20, and I feel like I was more mature than a lot of them like those

who were fresh out of high school. There was one 17-year-old who had a baby. She had so many problems and she tried to commit suicide. I was wondering how in the world she would manage her life as a Marine if she couldn't even handle Boot Camp. Back when I went through, I think the process weeded out the weakest females who shouldn't be there and those who wouldn't train. We actually had one handcuffed and taken away by the MPs. We were out drilling and the Drill Instructors stopped us and made us watch while they took her away. Of course, looking back, I'm sure they did nothing more than discharge her, but at the time that was pretty intimidating stuff.

Sgt Justine went through Boot Camp twice due to an injury, finally graduating in 1999. She agrees that the recruit training process creates a quality product while eliminating those who should be eliminated:

> There were always a few who didn't give a shit about anyone but themselves, and everyone including the DIs knew who they were and they were ostracized or IPT'ed [Individual Physical Training] until they came around. But mostly, it was good with teamwork and each person contributing her skills for the good of the platoon. The ones that were overly needy were weeded out very early on. They just couldn't hack it and did the seabag drag to Seps [Separations Platoon]. Mostly, we were glad they were gone because they tended to be the ones to get the whole platoon into trouble or wasted the efforts of the girls who did try to help them.

Captain Kami remembers an incredibly high attrition rate in her OCS class in 1998:

> At OCS, we dropped over 50% of our [female] platoon. We started with 64, I think, and graduated about 30. From the very beginning, a few of us said, "If *she*, whoever *she* was at the time, graduates, I will refuse my commission." Thank *God* we didn't have to own up to that oath because I don't know if we could have done it. By that point, the commission meant everything to us from what we'd been through to what it represented to us for the future. Honestly, I think that for the most part the ones that should have been dropped were dropped. There were a handful of girls that were marginal, those that did the minimum to get by but didn't cause any problems. I guess you have to have a low end in everything, even the Marines. [Laughs]

Mental Management: That Which does not Kill Us Really Does Make Us Stronger[37]

The mental and physical intensity of Marine Corps Boot Camp and Officer Candidate School is unrivaled as initial military training. So many experiences at Parris Island and Quantico are completely beyond the realm of the everyday that the challenges initially appear overwhelming. Over that twelve-week period, though, major transformations take place. All female Marines interviewed expressed at least one activity or event that was terrifying. For SSgt Tomasa, the gas chamber at Parris Island was her nemesis:

> The hardest thing at Boot Camp was the gas chamber. I *hated* the damn gas chamber! I had no idea what it even was. When I got in that room and they said to take the mask off, I was like, "Excuse me??" They were like, "Just calm down," and I was like, "Naw. This just ain't *normal*!" [Laughs] I actually ran over my Senior Drill Instructor and escaped the first time, but they made me go back in and we all suffered until I got myself together and we completed the training exercise. I did it, though. To this day, though, I feel bad that because I freaked out, the rest of the group had to stay there and wait for me to get my head together. There were no hard feelings, though. Damn, I *hate* the gas chamber!

The thought of OCS hikes terrified Captain Sophia:

> The humps, the hikes…I was terrified because I knew that I had spent the last three years trying to prepare for it, but I knew I was going to struggle anyway. My maiden name was Hurtz and when I started to fall out of those humps, those male and female drill instructors ate me for lunch, males and females alike. I broke my nose at OCS on one of those hikes. I tripped and I fell. Just like a little turtle, I had that huge pack on my back and did a face plant and broke my nose. It actually got me plenty of "cool points" and a lot of respect from my male counterparts because I was bleeding and nasty and we only had a mile to go. I was like, "Screw it! I have made it through this 10-mile hike and I am not quitting now." My driving force at OCS was, "I am not doing this again. I am doing it one time. I am not paying back my tuition. I am gonna be a Marine."

Cpl Sierra thought she was prepared for Boot Camp, but the

running was more than she anticipated, even to the point of being physically ill until the PT portion of the day had been endured:

> Physically, even emotionally, I felt prepared [for Boot Camp]. There were a lot of challenging moments, definitely. I felt like… well, I had danced all my life, so I felt prepared enough. We did the one-and-a-half mile then, and I had never run before in my life. But, I did it in, like, 15-something. So, they said, "Well, you did it in the 15 minutes, so you should be okay." So, I thought I was prepared, but I wasn't really prepared for the running. But, I always told myself that I had a child, so since I had gotten through childbirth, I could handle whatever they got to dish out. So, I think that's probably what got me through, 'cause I came through uninjured, everything. Every day, though, I was sick, sometimes even physically sick, anticipating the PT run. After that, though, I was good for the rest of the day. Then the next day would start all over again. It was brutal.

MGySgt Geneva describes her experience having been selected for Warrant Officer, but failing at TBS because of her mentality:

> The most challenging thing for me was when I went to TBS. I was actually a Warrant Officer. The Marine Corps as a whole has always allowed me to attend all of these professional schools and be successful. Then, though, I go to this school and it is totally geared toward the infantry. I know that every Marine is first a rifleman, but I'm saying that the school itself felt worse than Boot Camp. I applied as an 01 [Administration] and was selected as an 01, and went to TBS in February. I failed academically. I did not struggle physically at all with the runs and the humps and the swimming; it was the testing part. I think I just set myself up for failure, and I don't blame anyone but myself. Because I was an 01 SSgt with 13 years in becoming a 01 WO, I just refused to conform to the grunt mentality. That was not what I had been taught as a Marine and as a professional, to allow myself to be treated that way. We were treated like we were PFCs again. I was all motivated because I had had such an incredible career to that point with meritorious promotions and great success, but I hit a wall. That's just the way the program is, though. They're not going to change it for me. I think as a younger, more junior Marine, I would have had a chance. I was already too entrenched in a professional

leadership mentality, and I felt like TBS was just taking me backwards and in directions I wasn't comfortable going.

So many aspects of initial training are beyond the comprehensive scope for many young women. The aching loneliness of being away from home, physical and mental challenges, body management issues, and yelling Drill Instructors combine to create an intensely stressful environment. Nietzsche, however, was correct that "[o]nly a very few people can be independent: it is the prerogative of the strong,"[38] and the Marine Corps has banked on that fact for well over 200 years.

Female Drill Instructors

The strength of the female recruit training program is determined by the strength of its cadre of trainers and mentors. Female Drill Instructors are unlike any other women on the planet. Whether considered to be nurturing or demonic, they are respected as the superior leaders they are. Every female Marine has an opinion. LCpl Cheri, for example, describes her Boot Camp experience in 2005:

> Boot Camp was awesome. My senior Drill Instructor and my Kill Hat were so incredible. Even now, I will e-mail my senior and I refer to her as my angel. My Kill, I can best describe as a coconut. On the outside she's hard and rough, but on the inside, she's sweet and nourishing to the mind and soul. This is, I think, one of the best ways to describe what the "perfect" female Marine is or should be: a person striking that perfect balance between strength and softness…Although harsh in previously unimaginable ways, Boot Camp was about love and caring and nurturing. The same is true for the males, but they'll never admit it. It's about the Marine Corps investing in you, as a person, and bringing you into the family. It's about building trust. It's so much like a family. Sometimes Mom's a hard-ass. Sometimes it's Dad. Sometimes your older sister or brother chews you out for doing something dumb. In my experience, there's not really too much of a stretch from my civilian family to my Marine Corps family. They all had different tactics, and we kinda knew what to expect from each one.

Sgt Rocio relates her experience in 2001:

There were no surprises until I got to Parris Island and people were in my face screaming. The experience was definitely more mentally challenging for me than physically. I wasn't used to the constant criticism. It was amazing that the Drill Instructors could find fault with every single thing I did. Nothing, no screw up however minor, was overlooked. And, I think, if I hadn't screwed up, they'd make up something! It blew me away for awhile…until I realized that it wasn't all about *me*. Through their "abuse," they were teaching me things about myself, about human nature, and about where I really fit into a group. This is something that really made me think because, where I came from, we had gangs, some very violent gangs. I started to recognize behaviors that sucked people in, coercion methods, and playing one person against another, and the loyalty games. I began to see how the *vatos* in the neighborhood used those same methods to seduce with their power and then hurt people with their angry, pitying control. These incredible females, though, these incredible female Marines, used the same tactics to build people up. They used them to empower us to make positive decisions and to work as a team and to bring everyone along with us. We didn't leave the weakest among us behind. We found her strengths and brought her along, too. However challenging it was, it was worth it. It was truly emancipating!

Male Drill Instructors

Although not a constant presence for female recruits, male Drill Instructors occasionally make an appearance, which is sometimes positive and sometimes negative. Sgt Tabitha describes an incident on the confidence course:

I wanted so badly to throw something at one of the male Drill Instructors one day. We were doing the inverted wall and I knew it was all technique, but I also knew that a lot of those girls weren't going to be able to do it without help. So, I helped all of my team over first and I was the last one to get over and I hear this guy say, "Oh, great. This is gonna take all day." I was thinking, "Oh, no, I did *not* just hear you say that." I flipped myself over that wall like it was nothing and I hear him say, "Holy shit!" I was like, yeah. Hell, yeah. And don't talk trash about females not being able to get over the wall 'cause I've seen a bunch of your sorry male recruits not get over the friggin' wall either. And, they couldn't jump that stupid rope over the water thing either. I couldn't believe that because how could he talk trash when we were all watching the males fuck up?

In her position of leadership on the Island, Captain Joy actively promotes the interaction of male and female Drill Instructors in front of recruits of both sexes:

One thing that I make a point of is when we track, which is one platoon of females on the same training schedule as a male company, I make sure that the male and female recruits see male and female Drill Instructors and company commander working together, talking together, planning together. When appropriate, I encourage female leadership and reinforcement of the "team" concept so that those male recruits see females being competent and proficient and in positions of leadership right there with their male Drill Instructors. The interaction should all be the same. It should not be male or female, but all *Marine*. I think San Diego Marines can be a problem only because they haven't witnessed females doing the same things they've done. They've never seen or heard a female Drill Instructor in all her volatile glory. I think it's unfortunate because when they are out in the Fleet, they just might be working for or with females, and they need to understand that we've gone though the exact same things they've gone through. There is no difference anymore. That's where respect can grow, in understanding that we're all in this together and we have to be able to trust each other.

CWO3 Phoebe, on the Drill Field from 1996 to 1998, relates why male Marine Drill Instructors are *expected* to be dominant on the Drill Field and how she handled the situation:

We don't need *men* to come in and teach us the "masculine" stuff about the Marine Corps. I had nuns as teachers who were scarier than *any* male Marine I've ever met! On the parade deck, if the guys barked orders at those females to make those rifles move fast, then the female Drill Instructors needed to bark louder or with more teeth to get those rifle moving faster. This relinquishing of power to men because they have a penis is just wrong. Their voices aren't louder. You just let them be louder. They're not necessarily wiser, you just let them believe that they are. Why give up your power to them? They just take it and run, and you get nothing at all but their kids and their laundry and their bills, etc. etc. etc. Why? I really don't see why we ever put up with this.

In the intimate surroundings afforded by the Marine Corps, females often become more aware of the limitations of their male counterparts. Some, such as Captain Joy, are in a position to "fix" relations between the sexes. Others just take the little victories wherever they can get them. Regardless, the fact remains that not all males are superhero warrior-types and not all females are weak and demure. Basic biological differences and the expectations for each continue to play out every day in the Marine Corps. Although some writers surmise that females weaken a military force,[39] there are other variables impacting the performance of both sexes that warrant discussion.

Changes in Intensity & Perhaps Quality

Historically, the Marine Corps has used harsh tactics to produce the desired, indeed required, honed product, United States Marines. SgtMaj Denise Kreuser admits that some call it brainwashing, but asserts that the goal is to instill in young people honor, courage, and commitment and the credo that Marines never lie, cheat, or steal and to do what's right for the right reasons.[40] While recruit training has changed positively in recent decades to bring females up to speed, literally, with equivalent training to that of males, many female Marines since the AVF generation believe that other aspects of the experience have become too "soft" to produce the superior Marines, both male and female. This is reminiscent of the concerns expressed by Moskos and Meštrović about the future of a postmodern and postemotional military devoid of authentic emotional and professional commitment.

SgtMaj Brittany identifies a small, but significant contradictory change to the Boot Camp experience as the requirements for females became more stringent between 1981 and the early 1990s when she returned to Parris Island as a Drill Instructor:

> As a female back then, like in Boot Camp, we weren't exposed to so much that the females do today. We fam fired the rifle. We had a class on patrolling and did land nav and that, but it was nothing even remotely close to what the males did and what the females are doing today. It was another world. It's funny, though, because when I was a recruit, we did live in shelter halves. When I was a Drill Instructor, though, female recruits were staying in a hard-sided building. I thought that was an odd reversal.

The "odd reversal" Brittany describes is that females of the

AVF generation who participated in extremely limited combat and weapons training went to the field at Elliot's Beach and slept on the ground in tents, with each recruit providing "half" of the shelter that was constructed by the two of them. Currently, however, during field training at recruit training, females sleep in buildings with racks [beds] and head [bathroom] facilities. Thus, while the females' training intensity has caught up to the males', their field amenities have become more "luxurious" than those of previous generations of females due to modern "feminine hygiene" concerns. An "odd reversal," indeed, that exacerbates males' allegations of female favoritism while radical feminists cry "victory" in women's alleged sameness with men.

Also speaking to the modern Boot Camp experience, although she is convinced that the majority of Marines coming out of Boot Camp are hard chargers, Former Drill Instructor 1stSgt Katrina shares her displeasure in the contemporary training environment:

> Things sure have changed. The recruits can now tell you, "One arm's distance, Drill Instructor." Why do they know that? Who taught them that? Drill Instructors are not allowed to say anything harsh to the recruits anymore. It's a babysitting camp now, so they're not getting that sense of authority, that sense of trusting their leadership anymore. When I was there, I told it like it was because I had screwed up every way to Sunday. Now, though, they look at Drill Instructors like they're something they just have to endure for three months to get out in the Fleet and get a paycheck. They watch Boot Camp on TV, so they know what to expect. All the new rules protect the recruits' "sensitivity" and "self-esteem." This is for males and females, so no one can say "It's the *women* screwing everything up at recruit training." So, someone like me, now as a 1stSgt, I have to be the bad guy when there have been leadership failures all along and they are primarily the PC [politically correct] environment that we now have to train and live in. I can't *fix* all these problems. I just live day-to-day and do the best I can with what arrives at my hatch [door]. Recruits are leaving Parris Island the same way they showed up. There's not as much transformation as there used to be. I see it eroding year after year. Drill Instructors now have to follow the RTO [Recruit Training Order] to the letter, and they worry more about an allegation or a congregant or their career than just doing the job of training recruits. It's a paradox that we can't escape from, and it's the Marine Corps and all the Marines in it that are suffering because of it.

MSgt Madeline, a Drill Instructor from 1998 to 2001, identifies changes that she sees in young Marines today:

> The people coming into the Corps are definitely changing, maybe not in motive, but in attitude. I see the kids coming in and they are smarter, technology-wise, than in the past. I think they're accustomed to living in a "microwave society" where they want everything right now, but in the Marine Corps that's *not* the way things work. They want rewards, like promotions, *right now*. Heaven forbid they're a corporal for more than a year! They just don't seem to understand that the rank structure involves a maturation process. We do volunteer work and in the past, we've always had a lot of volunteers. Now, though, the attitude seems to be, "What do I get out of it?" I see a lot of them who seem to never have been held responsible for anything before. Their backgrounds and upbringing seem to be a lot different than the past couple of generations. I don't want to say "Old Corps" or "New Corps" but there is definitely a difference in the young Marines that we have in the military now. Believe it or not, it's *not* difficult to meet the needs of all of these diverse expectations because, after all these decades, the Marine Corps remains one *thing*. The Corps isn't pandering to the perceived needs of potential recruits or candidates. It hasn't compromised yet, and I pray that it never will.

Cpl Jenna, a recruit in 2005 and now working on staff at Parris Island's Recruit Training Regiment, describes changes she has witnessed in Boot Camp within the past few years:

> I think the whole training process has gotten easier. I don't know if it's political or what, but Drill Instructors can't talk to recruits like they used to. The attitude is that they can't hurt the recruits' feelings and they have to be politically correct. I know this one Drill Instructor who's on her second tour and she hates it. She says that it's more like Girl Scout Camp than the Boot Camp that she went through. Even though it was less time, in actual training days, it was much more intense and much more transforming because of the sheer hardness and unreal-ness of the experience. I'm not saying recruit training is "bad" now or ineffective, but I think that the enemy that you might eventually face doesn't care if he hurts your feelings or not, and maybe we should be developing a hardness that was there in the past, but maybe isn't there so much anymore.

Another corporal, Alecia, also a recruit in 2005 and now stationed aboard the Depot, expresses frustration with the current system:

The point of Marine Boot Camp is to instill in a person the willingness to accept the way things are, to trust in your leaders and to do what they tell you to do. I'm a boot, so I don't know it all by any means, but it just seems like an exercise and the Drill Instructors work so hard; you can *see* that. But, when it comes down to it, they seem also to be spending a lot of time and energy on girls that just don't want to be here. Now, the guys are the same way. There are those that just constantly screw up. They don't want to be here and they make it obvious to everyone. But, as women, we stick out so much that, if there's an "issue" with a woman, it becomes this huge blow-out-of-proportion "problem" that everybody talks about and everybody thinks they need to "solve." I think the best way to solve it is to ship 'em home. Send 'em a bill for their training and see if there's another, better girl to take her place. I feel like the focus on the attrition rate is the wrong thing to focus on. We're here for a much bigger mission than to make numbers either, one, to get females to Boot Camp, or, two, to graduate "x" number of females from Boot Camp. If they keep graduating poor quality females from Boot Camp, yeah, it does make the whole Marine Corps look weak. The strong ones can work so hard and can only compensate so much but, in the end, we're stuck with the stereotypes from past generations of feminine weakness and a dependence upon the destiny of our biology.

Arriving at Boot Camp in 2004, Sgt Eileen describes her frustration when the system doesn't work:

The system didn't weed out the ones that I thought shouldn't be there. I am one of those girls that *had* to do well. I just suck it up and get the job done. If I felt sick, it didn't matter. I was out there giving 110% at PT. We had a handful of girls that were on light duty the entire time and still graduated with us. It's very frustrating when there are those girls that slipped by and you wonder how they can make it in the Fleet if they can't make it here. Sure, Boot Camp might be an extraordinary experience, but it's just the beginning of being a Marine. If you're a heat case in Boot Camp because you say the water tastes nasty and you won't hydrate yourself, what the hell are you going to do in Iraq?

These examples of initial training in the Marine Corps reinforce tenets of evolutionary psychology as well as sociological theory, especially Meštrović's identification of America as a Postemotional Society. Meštrović states that postemotionalism assumes that almost all aspects of contemporary social life revel in the useless, in luxury, and that contemporary indignation is also not linked to appropriate action, and is another luxury emotion.[41] Although female Marines display a unique determination to become United States Marines, it is from American society that the military draws its modern-day warriors, male and female. Thus, contrary to Brian Mitchell's and others' assertions that the U.S. military has become "feminized" by the presence of females, it is more likely that the entire *nation*, in its postmodern, postemotional state has become not more feminized, but neutered by apathy, ambivalence, indulgence, and indignation.[42]

The End Product

Even in expressing their concerns over the attitudes and behaviors of present-day Marines, many female Marines laud the many positive changes in initial training enacted to bring the level of initial Marine Corps knowledge and skills of females up to that of the males. The process still creates a unique individual: a United States Marine. Regardless of the generation of the Marine, all expressed an overwhelming sense of accomplishment upon completion of Boot Camp or Officer Candidate School. Sgt Ruby, for example, a self-professed "bow head," remembers the transformation of herself and her platoon into U.S. Marines in 2001:

> Boot Camp was awesome. The motivation, the discipline, just gave me a totally different perspective on my abilities than I could even have imagined. Before, I was just selfish and self-centered, doing what I knew everyone in my little world expected me to do. I had no sense of what was going on beyond what I knew and understood. Going to Boot Camp opened up entire new worlds that I didn't even know existed, and it was about the team. It was about bringing all of these totally random people and experiences and beliefs into one place to form something new and different and stronger than anything those other singular things were separately. We learned from each other. We helped each other. We drew upon a history, a proud heritage, and we were tasked to measure up to that ideal. It seemed crazy at first, but the Drill Instructors gave us

the tools and the skills and the knowledge to figure it all out. To stand on those yellow footprints, knowing that you've cast aside your entire life prior to that moment and to be so isolated within a group with so much pressure on you is something no one can understand until they've lived it. You don't have a clue where you future is taking you. None. It's frightening, but it's also liberating.

In spite of its flaws, Cpl Alecia agrees that the life-changing experience of Boot Camp is well worth it:

Boot Camp is one of those experiences that you're fully brought out of your comfort zone. Most Americans will never, ever experience that. When you have to shower with 70 other girls, it's crazy! You lose the inhibitions. When all the external stuff is stripped away, you realize that we're all more alike than we are different. You find ways to get along and you learn to appreciate the different talents and skills that other people have that you would never have imagined just passing them on the street. I'll never again look at someone and immediately make a judgment solely upon what I can see with my eyes.

SSgt Miranda sums up the entire experience succinctly:

Everybody has a role to play, and our basic training teaches us that. Not everyone can be the strongest or the fastest or the smartest. That's not what the Marine Corps is about. It's *never* been about that. I think so many people see this image on a commercial or a movie and they have this idea of what a Marine *is*. The image they see is really an amalgamation of *all* of the traits and principles that we all aspire to. What they see is the ideal type. The rest of us work hard every day to ensure that the ideal is maintained as such, a set standard proven in the past and leading the way into the future. The Army has that slogan, "An Army of One." That's bullshit! A fighting or peacekeeping force is made up of so much more than any one person. You don't see that in Marine Corps commercials or billboards. You might see a single Marine, but he or she is only put before you as the example. Our Corps has over two hundred years of examples to guide us. It's never been about "one" Marine. It's always been about the team, the *esprit* that binds us all together generation to generation, and the few exceptional people who are personally recognized for particularly heroic feats. We all aspire to that heroism, but we're human, too, so we do the best we can. Sometimes it's offering a few words of hard-knock-earned wisdom. Sometimes it's assisting a blind person across a busy intersection. Sometimes it's telling the truth when someone wants to hear a convenient lie. A hero might be right in front of you and you never even know it. These are the people Marines strive to be, and I am extremely proud to be but one of them.

To return to 2nd Lieutenant Santangelo's allegation that the Marine Corps sets up its females for failure through sex/gender segregated training, Captain Caroline speaks to the heart of the issue:

> Segregating recruit training by gender serves a very specific and valid purpose; however, it ends up being an unrealistic preparation for what women can expect in the Fleet. In theory, yes, the training is the same because the training schedule is the same. But, if the men are running faster at every point in the PT schedule, the females are falling behind already. If they're not interacting with males, they're not developing the very fundamental skills that women will need to develop in dealing with an organization that is almost wholly male. In the RTR, they try to evenly distribute infantry officers and SNCOs into the structure because infantry guys are "…better warriors, better leaders, blah, blah, blah…" So, what happens with 4th Battalion? Because females can't be in infantry, by default, then they can't provide this supposedly superior leadership to females. Fourth Battalion and the females it trains are inadequate from the start, and that's sad. Then, there's the snowball effect by which the male recruits learn infantry tactics almost by osmosis, living with these grunt DIs every day, but the females will never have that experience. It's a handicap from the very beginning and it just continues throughout a female's career.

Suffice it to say that the United States Marine Corps remains a unique organization training Marines to kill as well as to nurture, like the "strategic corporal,"[43] and often in extreme circumstances for each behavior. Marines are riflemen, first, and initial training tests a recruit's or candidate's mettle. Those who aspire to become Marines participate in indoctrination at Boot Camp or Officer Candidate School. There simply is no other way to secure the coveted Eagle, Globe, and Anchor emblem and title "United States Marine." Through those processes and throughout one's career, one continues to learn and develop as Marines, as leaders, as Citizens, learning to balance these two conflicting activities of destroying and nurturing.

In this, males and females are more alike than they are different. Through these co-evolutionary processes, kinship bonds form between like-minded individuals, those committed to being their best selves and best comrades, all maintaining the integrity of the Marine Corps in support of its mission.

The data herein confirm that objective, non-gendered, and non-negotiable standards for gender equivalence presently exist in the USMC in the form of core values, leadership traits and principles, and the Uniform Code of Military Justice (UCMJ). Training standards are at the highest level of equity in history, and Marines throughout the rank structure overwhelmingly are committed to the success of *all* Marines. In this endeavor, female Marines as well as male Marines will continue to coexist along a spectrum of acceptable performance. Female Marines who possess the requisite physical strength and endurance, the mental aptitude, and the leadership potential are marching down the road to full equivalency with their male peers and will ultimately succeed or fail as determined by their individual capabilities.[44] What must happen organizationally is the institution of the concepts of equivalency through kinship which will, in turn, foster increased empathy and camaraderie among all Marines.

Endnotes

1 Brownson, "Battle for Equivalency," 2.

2 Colette Dowling, *The Frailty Myth: Redefining the Physical Potential of Women and Girls* (New York, NY: Random House, Inc.): 259.

3 Seck, "Marines Delay Female Pullup Requirement Again".

4 Brownson, "Battle for Equivalency," 2.

5 Ibid., 15.

6 Sage Santangelo, "Fourteen Women Have Tried, and Failed, the Marines' Infantry Officer Course. Here's Why." *The Washington Post*, Opinion, http://www.washingtonpost.com/opinions/fourteen-women-have-tried-and-failed-the-marines-infantry-officer-course-heres-why/2014/03/28/24a83ea0-b145-11e3-a49e-76adc9210f19_story.html.

7 William Easter, "The Marine Corps PFT: Not Equal, Not Fair," unclassified, submitted to USMC, Command and Staff College, 20 February 2009.

8 United States Marine Corps Official Website > Home > Units > Recruit Training Regiment > 4th Recruit Training Battalion, Marine Corps Recruit Depot,

Easter Recruiting Region, Parris Island, Couth, Carolina, http://www.mcrdpi. marines.mil/Units/RecruitTrainingRegiment/4thRecruitTrainingBattalion. aspx

9   Marine Corps Order 6100.13, *Marine Corps Physical Fitness Program*, 1 August 2008.

10 Buss, *Evolution of Desire*, 26.

11 Veblen, *Theory of the Leisure Class*, 53.

12 See Peter M. Blau, *Exchange and Power* and Steven Goldberg, *The Inevitability of Patriarchy.* (New York, NY: Morrow, 1973).

13 For a contemporary example referencing previous studies, see Connie Brownson and Harold Dorton. "At the Margins of the Minors: Good Girls, Bad Girls, and Baseball Beyond the Big Leagues." *Popular Culture Review* 17 (2006): 67-78.

14 Brownson, "Battle for Equivalency," 21.

15 King, "Women Warriors," 6.

16 Brownson, "Rejecting Patriarchy," 2.

17 Brownson, "Battle for Equivalency," 21.

18 van Creveld, *Men, Women and War*, 227.

19 Regina F. Titunik, "The Myth of the Macho Military," *Polity* 40, no. 2 (April 2008): 147.

20 Carl von Clauswitz recognizes the element of chance in war, stating, "…in the whole range of human activities, war most closely resembles a game of cards." See Carl von Clausewitz, *On War*, ed. and trans. Michael Howard and Peter Paret. (Princeton: Princeton University Press, 1976).

21 Daniel D. Martin, "Organizational Approaches to Shame: Avowal, Management, and Contestation," *The Sociological Quarterly* 41, no. 1 (2000): 143-144.

22 Browne, *Co-ed Combat*, 259-260.

23 Antoine B. Hannoun, Anwar H. Nassar, Ihab M. Usta, Tony Zreik, and Antoine Abu Musa, "Effect of War on the Menstrual Cycle," *Obstetrics & Gynecology* 109, no. 4 (April 2007): 929.

24 James L. Brain, "Sex, Incest, and Death: Initiation Rites Reconsidered," *Current Anthropology* (1977): 198.

25 Ridley, *Red Queen*, 293-295.

26 Anthony Synnot, "Shame and Glory: A Sociology of Hair," *The British Journal of Sociology* 38 (1987): 382.

27 David M. Buss, "Strategies of Human Mating," *Psychological Topics* 15 no. 2 (2006): 243.

28 This section describes how a female Marine's agency is exerted within a framework of military regulations where a female does *not* have the option while in uniform of using her hair to gain power. In contrast, see Rose Weitz, "Women and Their Hair: Seeking Power through Resistance and Accommodation," *Gender and Society* 15 (2001): 667-686, for a discussion of females using their hair as a means to secure power.

29 Buss, *Evolution of Desire*, 53.

30 Holly Yeager, "Soldiering Ahead," *The Wilson Quarterly* 31 no. 3 (Summer 2007): 62 and Glamour Daze: A Vintage Fashion and Beauty Archive, http://glamourdaze.com/history-of-makeup/1940s.

31 Marine Corps Order P1020.34G w/Ch 1-4 of 31 Mar 03, *Marine Corps Uniform Regulations*, subparagraph 7.c(5).

32 See Stremlow, *A History of the Women Marines, 1946-1977*, vol. 2, (Washington DC: U.S. Government Printing Office, 1986), for an interesting discussion of uniform changes prior to the 1980s.

33 Ibid.

34 Ibid,, 100.

35 Ibid., 153.

36 King, "Women Warriors," 2.

37 Friedrich Nietzsche, *Twilight of the Idols, Or, How to Philosophize with a Hammer*, translated by Duncan Large (New York: Oxford University Press, 1998), 5.

38 Friedrich Nietzsche, *Beyond Good and Evil: Prelude to a Philosophy of the Future,* trans. and ed. Marion Faber (New York: Oxford University Press, 1998), 29.

39 See Browne's *Co-Ed Combat*, Mitchell's *Women in the Military*, and van Creveld's *Men, Women, and War*.

40 Larry Smith, *The Few and the Proud: Marine Corps Drill Instructors in Their Own Words* (New York, NY: W.W. Norton & Company, Inc., 2006): 218-219.

41 Meštrović, *Postemotional Society*, 26 and 57.

42 John Allen Williams, "Toward a Postmodern Military: The United States," in *The Postmodern Military: Armed Forces after the Cold War*, eds., Moskos, Charles C., John Allen Williams, and David R. Segal (New York, NY: Oxford University Press, 2000): 273-274.

43 Dayne E. Nix, "American Civil-Military Relations: Samuel P. Huntington and the Political Dimensions of Military Professionalism" *Naval War College Review* 65 no. 2 (Spring 2012): 94.

44 Brownson, "Rejecting Patriarchy," 6.

Chapter Four

Marine Combat Training (MCT), The Basic School (TBS) & Military Occupational Specialties (MOSs) Schools: Entering & Navigating the Marine Corps Work Force

After initial training, Marines' first step into the "real," occupational Marine Corps is the Military Occupational Specialty (MOS) school for enlisted Marines and The Basic School (TBS) for officers. This is where the day-to-day routine of garrison, field, and occupational life begins. Having passed the test to become a U.S. Marine, this is where one learns one's job. It is also where a woman must begin to manage a skewed sex-ratio occupational and social environment. Suddenly, the insulated structure of the initial training environment transitions into a workforce environment. A female Marine finds herself surrounded by male Marines, some brotherly, others hostile, a few predatory, and the remainder operating somewhere in between.

Depending on the MOS assigned, female Marines' experiences vary dramatically and often directly affect, positively or negatively, the course that a young enlisted Marine's career will take. For example, a female administrative clerk may have other women in her training environment and form a gender-specific social network, but a female engineer may find herself completely alone in a company of a hundred men. There may not even be another female Marine present at the training site. Female Marines must now perform physical training (PT) with the males, march with the males, and live with the males in a closed and isolated community. The standards for her performance and conduct, on and off duty, are daunting as she finds herself, literally, in a man's world.

The first generation of Women Marines experienced a solidarity based upon a shared social existence in an environment designed to maintain separateness between male and female Marines. Women Marines, prohibited from being married, lived together in the barracks as a separate but parallel organization to the males' Marine Corps. For example, Irma describes the camaraderie among her fellow women in 1944: "I had women friends in the Marine Corps, very good friends. There were about six of us who hung around together. We lived in the barracks and were constantly together. We did our laundry. These five and myself seemed to hit it off well, and we'd go to the beach and just hang around together."

Another World War II veteran, Sandra, describes the sense of *esprit de corps* she felt with her Marine sisters: "You always hear about *esprit de corps*. I had heard of it, too, but I think you can't really *know* what it means until you experience it. After about three days, the teamwork began, and the sense of being a band of sisters began to develop. It was an incredible feeling. There was no competitiveness among us, just a feeling of cooperation." Both ladies describe an environment in which women stepped in to do a job during wartime without the expectation that their roles would be any more than those strictly dictated by the establishment, both civilian and military. For example, Irma further states: "We were called US Marines, Women's Reserve. We didn't have any other name. We were Marines as far as everyone was concerned. The men seemed to respect us. They helped us if we needed help, like when we were working on trucks and we couldn't handle the big tires, so one of the men would come over and take off the tires for us."

During this period, there was no expectation that women would command men, no expectation that a woman could and should do a job as well as a man. It must be acknowledged that the experience in the "Free a Man to Fight" and contemporary generations' Marine Corps experience are quite alien to each other. While reaping many of the benefits of being "Marines," WWII females lived a relatively sheltered existence, commanded by females, maintained in a safe and supportive environment, and then whisked back to their civilian lives when the job of winning the war was done. Although committed to a common goal, men and women were largely unknown to each other as peers.[1]

In contrast to the Free a Man to Fight generation, as females became more integrated into the Marine Corps, particularly during the pivotal All-Volunteer Force era, their role evolved as well. Females were now expected to earn rank competitively with men and acquire leadership skills equal to those of the males. Instead of living in enclaves with other women, this generation of female Marines was expected to go their own way in the Marine Corps and carve a broader, more "military" experience similar to that of the males from a common foundation, Boot Camp or OCS. As a mere 7% of the Marine Corps population serving around the globe at any point in time, the result is that female Marines often experience alienation from their male peers and even from the females within their own chain of command.

Purgatory as the Final Purification of the Elect[2]

In 1989, Christine Williams asserted, "It is not unusual for military women to feel as though they've been 'thrown to the wolves' when, after completing training, they arrive at their first duty station: Their experience in basic training does not prepare them for a work environment in which they may be the only female."[3] There are actually three clarifications that must be made from this one seemingly simple sentence: 1) the next step after initial training is now Marine Combat Training (MCT) for enlisted Marines or The Basic School (TBS) for officers, which is actually an infantry-based purgatory of sorts, a more complex training and testing ground, between Boot Camp/OCS and the "real" Marine Corps; 2) more of a problem than being the "only female" is likely the fact that female Marines must now compete with males in an extremely competitive and physical environment; and, finally, 3) the issue of possibly being the "only female" surrounded by males creates social and sexual tensions. These realities comprise the female Marines' world and the new challenges she must navigate.

To address these issues, Sgt Eileen describes the difference between Boot Camp and the real Marine Corps, and how a female Marine must learn to police herself:

> Boot Camp is such a structured environment and you have no contact with males or the outside world other than through the mail. There's no junk food, no beer, no TV. There's nothing but

the focus on what you're doing right at that moment. It's not that you're not yourself there, but you don't have those distractions or temptations or opportunities that you have in the outside world to lure you into making bad choices. You make a bad choice in Boot Camp and you're on the quarterdeck. You make a bad choice out in the Fleet and you're pregnant or in jail or popping positive [on a urinalysis]. I guess what I am trying to say is that there is limited opportunity to really police yourself in Boot Camp, but in the Fleet you have all kinds of personal management issues to deal with as a female.

Earning the title "United States Marine" and wearing the coveted Marine Corps emblem is but the first step in a long journey to actually making things work, personally and occupationally, in the Marine Corps. From Boot Camp and OCS, female Marines embark on yet another phase of training that challenges them yet more stringently in the areas of physical competence, emotional stability, and moral virtue. Without the watchful eyes and barking voices of their Drill Instructors, new Marines male and female, now live and train together with only Staff Non-commissioned Officer (SNCO) or Non-commissioned Officer (NCO) guidance, training schedules, and duty rosters structuring and limiting their behavior. This environment offers opportunity for a young female, in particular, to establish her Marine Corps persona, ideally one of commitment, competence, and trustworthiness, a persona that will precede as well as follow her throughout her Marine Corps career.

At MCT and TBS after the sequestration of Boot Camp and OCS, the temptations for indulgence abound, potentially damaging one's reputation, career, and physical health. The first obstacle to negotiate is the overwhelming presence of desirous men everywhere. Although flattering, submission to the attention and ensuing stereotypes compromise camaraderie, a shared sense of mission, leadership, and respect.[4] The numerous obstacles to personal and professional success may seem insurmountable; however, a female Marine must establish the framework through which she may best nurture her physical/sexual, occupational, and leadership roles.[5]

Boys & Girls Together Again

A new female checking into a Marine Corps unit, whether overseas or stateside, can be disruptive to unit and even base operations.

Often referring to themselves and other new females as "fresh meat," female Marines recognize that their arrival at a new unit, especially that first unit or school, can be traumatic. Staff Sergeant Dixie stated it succinctly, "You're in the fishbowl. You're at the mercy of the system. You can't hide. You just show up and get it over with. The up-side is that you only check in there *once*." It is at that point that a female Marine makes that first impression. GySgt Claire, describes the experience and offers advice for neophytes:

> When I checked into my school, guys came out of the woodwork. They were hanging over balconies. They were trying to help us with our luggage. We were trying to check in and we were being invited to parties. That was the game that's played over and over. It happens every time the fresh meat arrives. That's the time a female needs to start establishing those boundaries because that's where she starts to define herself in the Marine Corps. She needs to conduct herself as a lady and let those men know that she means business and that she's not going to take their crap or play their games.

Although generations apart, Sandra, a World War II female Marine describes a similar experience in the late 1940s, "I remember getting off that cattle car that day at North Island. All of these men there were curious, of course, and they were all standing there, watching us like we were something exotic that they'd never seen before. [Laughs] It was a little awesome, but we all fit right in and not a one of us that I know of never had any problem."

Another WWII veteran, Darcy, relates her experience with a "player" at the end of the war, stating "There was this guy at a dance in Hawaii and he wanted to dance with me, so we did. And the band director said, "We just found this wedding band that belongs to one of you fellers." It was the one I was dancing with. [Laughs] He was working on me, see? Overall, though, they were very polite. They had great respect for us; that was the only disrespect I ever experienced."

Roughly 35 years later, the environment described by MGySgt Geneva at Camp Del Mar on Camp Pendleton, California, was not as respectful:

> It was tough leaving Boot Camp and entering the Marine Corps. I remember getting off the bus at Camp Del Mar and I was the only

one standing on the parade deck in my Alphas with my seabag. I remember the males coming to the windows and doing catcalls. I was 22 and I was thinking how immature that was. I mean, if you want to talk to me, just come over and talk to me. That was the *worst* way to try to get my attention because it was so rude and disrespectful for them to think that just because I have tits and a hot ass and knowing that I'm single that I'm the fresh meat on the block. I didn't have a problem with that because I thought that I could talk to these individuals after I got settled in, and I did. I'd tell them, "Look, you know what? You don't have to whistle at me. You *can* approach me and ask me my name and we can speak to each other." I think it was better for me to have been older 'cause I was looking at these guys like they were idiots because they were acting like they were 15- or 16-year-olds and back in high school.

1stSgt Katrina sees the pattern repeated ten years later in the 1990s:

What the males were doing with the women really didn't affect me because I wouldn't let it, but I *saw* it and it was so obvious and predictable. The new meat would arrive and the other girls would be forgotten as the men started to pose for this new girl. It was crazy! They were all trying to be the first one to get her. As I've come up in the Marine Corps and having talked to so many of these young women, they tell me that they're just not used to the attention. Like, if they come from a small town where everybody knew everybody, they come into this environment where so many men treat you like you're a goddess. You don't have to be pretty. You don't have to be rich or smart or even sexy. From the lowliest PFC to the SNCOs to the officers, *if they're not smart*, you're getting *all* the attention. It's so hard not to eat it up! Nobody ever takes them aside and tells them the way they need to behave in this situation. It can't be politically correct. It can't be done by an EO rep. It can't be done by the chaplain. You can't use fancy language. No. You have to get on their level and make it *real* for them. You have to tell them that these men just want to get laid. They don't love you. They're lonely and you're the newest, or the only, piece of ass they've seen in a long time. These girls need to be taught how to keep it in perspective.

Captain Kai identifies the two primary concerns of males when a female arrives at TBS, her competence as a Marine and her sexuality as a female, not necessarily in that order:

I went through TBS in 2001. The men are sizing you up as soon as you arrive. They're wondering if you can run as fast as they can, if you can hump as far as they can, if you can carry as much weight as they can, and if you're going to want them to carry your load. If you walk in there able to carry your weight, they'll love you immediately. You will deal with sexual politics. There will be some who are trying to sleep with you and, if you say "no" there will be issues to deal with there. It's just always a fact no matter where you go... There's definitely a lot of sexual politics at TBS. I joke with my friends now that I wasted TBS because I actually had a long-distance boyfriend that I remained faithful to. You're gonna get labeled a slut whether or not you sleep around so, in retrospect, I should have just done it and enjoyed myself. I wasted the buffet, standing there and looking, drooling sometimes, but not sampling. My dad even told me that before I went to TBS. He said, "You better watch yourself. It's a candy store." It totally was, but I blew it. [Laughs]

Captain Jordan describes how the same scenario plays out over and over in the course of a female Marine's career:

One Marine might meet a new female when she's checking in and, regardless of what happened in the exchange, all kinds of crazy impressions and dialog and all kinds of other things emerge. I swear I don't know how a guy can turn a female saying, "Thank you," into, "I love you." The leap is so easy for them, though. [6] [Laughs] I see it over and over again and it never seems to change. New Marines in a unit...same crazy behaviors.

GySgt Hailey astutely identifies this behavior as "mating" behavior, which clearly supports the hypothesis that the Marine Corps, as a microcosm, represents Buss' theory that human sexuality and mating form the very foundations of human motivation and behavior. Hailey, however, compares male Marines to turkeys:

As far as mating rituals, in many fields in the Marine Corps, when a new female checks into the unit in her Alphas and pumps, you can see a pecking order emerge with the men. Like, the male SNCOs will "drop by" the S-1 to check her out and then the NCOs, roughly speaking of course, then on down the line. If she's a real looker, they compete to take her around to check in and I guarantee you that it won't be a PFC who "wins" that honor. Un-uh. That ain't happening. It will likely be a SNCO or senior NCO [with a vehicle]. If she's beautiful and single, of course, she'll create quite a buzz. It's like watching turkeys during mating season, puffing up and shaking their feathers. She'll have her pick of any one. [Laughs] If she's attractive and married, though, you can see the disappointment and if word spreads around fast enough, there won't even be that many guys "dropping by" to see her. If she's wearing a ring, she's pretty much off limits until one or two guys actually fish information out of her about how serious her relationship is or whatever. If she's ugly, the desperate guys will pursue her, but not the "quality" ones. [Laughs] I've seen this happen over and over, and as the years have gone by this is one ritual that never changes.

From an evolutionary psychology perspective, this situation is identified as the Coolidge effect, "the tendency of males to be sexually rearoused upon the presentation of novel females," which is a wide spread mammalian trait and is supported by cross-cultural human research.[7] Similarly, Kanter determined that in the presence of token women, men exaggerated displays of aggression and potency: instances of sexual innuendo, aggressive sexual teasing, and prowess-oriented "war stories." When one or two women were present, the men's behavior involved showing off, telling stories in which masculine prowess accounted for personal, sexual, or business success.[8] Likely a heightened response in the Marine Corps environment because of the skewed ratio of females to males, it is evidenced in this research that this mating dance endures over time, at least 60+ years, in much the same manner but framed by the social constraints of the era in which it occurs. These social behaviors stem from mating behaviors in which women possess the power to establish the ground rules. Dovetailing on the discussion of women's prerogatives in the previous chapter, women shun men who are easily dominated by other men or those who fail to command the respect of the group.[9]

Queen Bees & Mate Guarders?

Female Marines also engage in posturing with other females; initial encounters range from welcoming to openly hostile. Sergeant

Eileen describes her specific experience in 2005 and how her arrival at MOS school stirred the interest of the males and the contempt of the females already present:

> When I arrived, it was like bees to honey. Those guys were all over me. I was 19, a PFC, brand new. I knew that I did not want anything to do with that scene. I isolated myself a whole lot. We had a female corporal and one other female and me. They were there and I was the new girl, so they weren't interested in reaching out to me. I think it was female competition. I was new. I was in shape. I was getting all the attention for no other reason than because I was the fresh meat, and they responded to that with negative attitudes, which is fine. I didn't let it bother me.

Similarly, Captain Jordan describes the females' reception to her in Okinawa and her decision to remain distant from their behavior:

> In Okinawa, I was mentored more by males than by females. I really didn't encounter too many women there but at TSB, the red patchers, there were a lot of female lieutenants and they were *very* competitive and territorial. I was not too much senior, but I had already done a tour in Okinawa, so had been a few places and done a few things. They were always skeptical about new Marines, especially females. They weren't so much clique-ish as they were pack-ish. By that, I mean that they weren't really even friendly or loyal to each other except when they ganged up *against* someone else. That was pretty amazing. While I was there, I just did my own thing and didn't play any of their games.

GySgt Claire compares the situation of young females in MOS school fighting over males to the similar situation in high school:

> At cook's school was where all the personalities really started coming out. In Boot Camp, you have the quiet ones or you have the ones who will get you in trouble, but there's not a lot of room for free thought or free behavior. At school, though, it almost seemed to flip-flop and the quiet ones became the ho's. It was amazing to see those girls always at each other over the guys. I was older, so I stayed out of it and just watched these little 17-year-old girls having their high school cat fights. I was like, "Damn. There's 20 other guys. Go find your own. Why y'all fighting over this brother?"

Here I am, thinking I'm an adult doing an adult job, getting paid to defend my country, and we shouldn't be having all that crap here.

LCpl Sadie describes territorial behavior of females in the barracks, stating, "What's interesting is that the other females, before they got to know me, considered me a barracks whore, *not* because I'm a slut but because I was getting attention from the guys that *they* want."

It is ironic, or perhaps appropriate, that Sergeant Eileen uses a bee analogy to describe her experience. The behavior she, Captain Jordan, GySgt Claire, and LCpl Sadie describe is a manifestation of the Queen Bee Syndrome, a concept originally proposed in 1974[10] receiving recent attention in the social sciences literature. In essence, women are particularly inclined to experience competitive or collective threat in response to agentic female colleagues. These women may have difficulty working cooperatively with females they perceive as threats to the status quo they have established and strive to maintain. This could be particularly true in "winner-take-all" climates wherein valued rewards, such as inclusion in prestigious groups are allocated to only a few.[11] Initially, this seems unlikely in the Marine Corps since the inclusion of all women, a mere 7% of the entire population, and a codified hierarchy via the rank structure seems reasonable to accommodate. However, when one considers that the Marine Corps is already a "prestigious group," inter-group competition by females for status among the 93% male population likely results in the cattiness described above.

Similarly, Buss asserts that feminist theory portrays men as being united with all other men in their common purpose of oppressing women, but the evolution of human mating suggests that men and women compete primarily against members of their own sex. Women compete with each other for access to high-status men, have sex with other women's husbands, and lure men away from their wives. Women slander and denigrate their rivals. Women and men are both victims of the sexual strategies of their own sex, so can hardly be united with members of their own sex for some common goal.[12] This is particularly true in the mating years. GySgt Hailey attributes the competitiveness among young female Marines to immaturity and insecurity of their place in the big picture:

For the first few months, I felt the competitiveness of the other females. My attitude was that we're the fewest of the few, and we need to learn how to work together instead of being at odds. I notice that with the other younger females. I see that the longer females are in the Marine Corps, the more cohesive we become and we're more willing to put aside our differences and focus on our similarities. We develop that "Been there. Done that," mentality and we see that success doesn't necessarily come from besting each other. We've already carved out our niche. I think age and experience have a lot to do with the success of females in the Marine Corps.

Similarily, Captain Maya describes a transition from early competition and then, over time, recognizing the value of female friendship based upon sex/gender connection in the closed system of the Marine Corps:

It's funny because, as lieutenants, you feel like you're competing against all the other women, so you shut them out. You don't want to be associated with other women because you're trying so hard to just fit in. So, you think, the more women you hang out with, the less you'll fit in. You *don't* want to be associated with a weak woman 'cause, all of a sudden, you feel like you're lumped into that group. But, I will tell you that developing female relationships in the Marine Corps is valuable, because the further on, the more and more you need them. They're the only ones who can truly understand where you're coming from, you know, like deploying and leaving your baby.

These female Marines and many others whose comments corroborate but are not included here consistently describe female Marines as territorial, manipulative, and jealous of "their" men, sometimes extending to include all men in their platoon or company. Many also recognize that as female Marines move up through the ranks, often marrying and having children as they do so, the tensions between females are not as obvious or intense, and strong bonds can be formed with other females. This, too, can be explained from an evolutionary psychology perspective. Buss writes that from a female's perspective, as her direct reproductive value declines with age, her reproductive success becomes increasingly linked with nurturing her children, the vehicles by which her genes travel into the future.[13] Thus, their sexual competition decreases. Further, females in and out [already mated] of the mating game display quite different levels of attractiveness to the males around them, and they respond to others from within the roles they have assumed. It follows that females already mated and no longer competing for an acceptable mate seek out like-type females for companionship, advice, and mutual respect. To many of these wives and mothers, the younger generation of females is little more than bothersome, unless they cross the line of attempted mate-stealing as affirmed by Buss above. In this case, the issue becomes personal as well as potentially detrimental to the Marines' unit(s).

Wolves in Marines' Clothing?: Breakdowns in the Chain of Command

The transient nature of young females moving through the Marine Corps experience perhaps ignites sexual arousal in Marines due to novelty.[14] Further, humans tailor mating strategies to the contexts we encounter.[15] Men in positions of authority and power historically have created situations in which sexual coercion, enabled by female preference,[16] allows opportunities for abuse in the military. Females new to the Marine Corps encounter situations in which male leaders attempt to "take advantage" of them. Incidents at the military academies, the Tailhook Convention (1991), Aberdeen Proving Ground (1996), Abu Ghraib (2003), Lackland Air Force Base (2009), and the Ft. Hood prostitution ring (2013) are just a few contemporary examples of military male-female sexual power relations gone horribly awry. In all of these unfortunate situations, the media tend to focus on *incidents* but not the social *climates* of abuse.[17] For example, Senator Kirsten Gillibrand (D-NY) relentlessly aims to take down the alleged "rape culture" in the U.S. military which, according to radical feminists, means derailing the patriarchal status quo.[18] In contrast, the stance taken by the Rape, Abuse & Incest National Network (RAINN) rejects rape as a systemic problem. Rather, RAINN acknowledges rape as a conscious decision made by a small percentage of individuals within the community to commit violent crime.[19] Like RAINN, all female Marines interviewed who discussed this issue testify that a female Marine's personal conviction, chain of command, and available victim services sufficiently mitigate incidents of sexual coercion. Violent sexual crime is a separate act/issue and the two should not be lumped together. Both will be discussed herein.

CWO4 Renae describes her experience as not just a boy-girl problem, but one of predatory males in her immediate chain of command, men who she, as a young enlisted Marine, should be able to trust and who, as Marine leaders, are charged with taking care of their troops:

> The biggest problem back then at the MOS schools was not just the boy-girl-out-of-Boot-Camp problem, it was the Instructor Problem. I have had this discussion with so many of my peers, and it always seems to have been the same situation in the majority of MOS schools at the time. The male instructors, and they were almost *always* male, would view the females as prime targets for sexual favors. They were blunt about it. I had a couple of instructors that actually said to me, "If you don't give me what I want, you'll fail." My response was, "Okay, if this is the case, you really need to look for another job because I'll turn your ass in and I don't care how far I have to go to get you busted and/or kicked out." This was when there was no "sexual harassment" or "equal opportunity." This was me as one of a handful of females in a group of over 100 men saying, "No. You're *not* going to manipulate me like that." It wasn't a bluff. It was the truth. I wasn't going to let anyone treat me like that. They left me alone. It's funny because so many men pull that bravado shit and don't have the guts to see it through. Often times, they're the ones bluffing and all a woman has to do is stand up to them and they back off.

SSgt Robin describes a bizarre event regarding a female instructor behaving extremely badly at Cherry Point, North Carolina:

> Of course, fraternization being what it is and punishable under the UCMJ, was a big deal, but a bunch of us would go to Greenville and hang out and party. One night, we had a female instructor, a sergeant, who went with us and I thought it wasn't right, but it wasn't my place to say so. I was the designated driver because I wasn't 21 yet, so I was ringside to the whole thing. This female started making out with not one, but *two* of the female students that were there! Come to find out, she was also engaged to marry one of the male students at the school who had gotten her pregnant. It was crazy! This was my first experience outside of Boot Camp with another female Marine and it really disturbed me so I talked to my instructor, my sergeant, about it and told him what

all had gone on. Come to find out, she was well connected. The gunny came to me, though, and was talking about how difficult it is to do the right thing, but I stuck by what I had told my NCO. Ultimately, they kicked her out of the schoolhouse, but that was just so disturbing to me at an impressionable time in my career.

These behaviors follow Buss' evolutionary reasoning that almost all of human behavior can be traced to reproduction/sexual imperatives employed within social and legal constraints. Mating behaviors of males and females in this microcosm of the Marine Corps result in the *female* choosing and dismissing potential mates at will, especially since the ratio of males to females is so skewed. The constant movement of Marines from one duty station replicates the Coolidge Effect and Queen Bees continue to defend their status quo hives. These sexual/mating patterns of Marines remain consistent over time and space at least as evidenced during and since World War II.

The Challenges of Physical Training

As described in the previous chapter, numerous female Marines espouse the belief that Boot Camp and OCS, prior to the implementation of significant changes in 1997, patently set female Marines up for failure in the Fleet Marine Force (FMF). Although they received similar training to that of their male counterparts, such as academics and swim qualification with rifle qualification added in 1985, they did not receive training in critical areas that potentially equalize the training standards for males and females in the Marine Corps to garner a sense of "fairness" between the sexes. Prior to the All-Volunteer Force, equity of training was not an issue; male Marines performed "manly" training and women Marines always did less and different, and conducted themselves as ladies first and foremost.

The Free a Man to Fight generation went through the motions of physical training first at Camp Lejeune and then as "regulars" at Parris Island beginning in February 1949. Calisthenics have always been part of the female Marine's physical regimen, but sometimes playing volleyball or riding bicycles was considered "PT," while officer candidates "worked out" by sailing on the Potomac.[20] Boot Camp lasted six weeks for the women, which has since been expanded incrementally over the past two decades to equal the males' twelve weeks. Also during that time, as females became more completely integrated into the All-Volunteer Force Marine Corps, dramatic changes in military occupation access, uniform accommodations, and physical training and testing standards have brought females closer to that elusive goal of "equality" with the males.

Achieving success within the physical standards historically has and continues to challenge females. Success in physical requirements correspondingly impacts females' leadership and overall success within the military organization.[21] For example, SSgt Lucy describes how difficult it was for her leaving Boot Camp in 1982 where running a mile-and-a-half was the mandated PFT run, but then being required to PT with males in the Fleet:

> It was hard being with a group of men who were all trying to be studly when we had been running at a pace in Boot Camp with only women. When we got there [MOS school], we were used to in Boot Camp, putting your weakest runner in front because you didn't want to leave anybody behind. So, the weakest person up front, in essence, sets the pace and in Boot Camp, it was rare that we would go at a faster pace, only like if the Series Commander ran us. Then, it was faster because she was usually a young officer and ran like a rabbit compared to the rest of us. Once we were with the men, we figured out quickly that regardless of where we started, they would leave us in the dust and not even glance back. This wasn't Boot Camp anymore. So, we basically just jumped in there and did what the men do. You find a way to get it done.

She continues to speak more to the issue of "equality":

> We were all young people just fresh out of Boot Camp so we really didn't have a sense of responsibility. We had basic Marine Corps manners, if you will, but we were just young kids. We did what we were supposed to do, but at night we hung out together and it was really quite normal. We, the women, knew at the offset that we were not "equal" to the men. The men let us know right away. "You women didn't stay in Boot Camp as long as we did." "You didn't do as much as we did." "We're bigger." "We're stronger." "We're faster." It was that kind of thing and made very clear. I guess we knew our place and didn't let it bother us much.

In direct contrast, SSgt Robin describes the importance of exceeding physical training standards almost 20 years later in 1999:

At the schoolhouse, there were maybe four females out of maybe 100 students, total. When I first checked in, we had to run a PFT. I volunteered to be a road guard and they said, "No," because they said I couldn't run fast enough. I told them I could, so they let me and I did. I was awesome and it was talked about. The males will boast about you when you're good enough. They don't particularly talk about their males that can run a 17-minute PFT, but they sure talked about me when I ran a 20-minute one. It was a big deal to them.

Similarly, SSgt Miranda found herself the last one standing during a training exercise, beginning with eight females:

There were about eight of us females who were in our bivouac [at Camp Hansen in Okinawa] and we did ambushes and pretty basic field training stuff. But, then, also included was an endurance course which included rappelling, hasty rappelling, and an entire jungle warfare course full of scenarios for a squad-sized unit to maneuver through. It is designed to last from two hours to twelve hours, depending on the squad's abilities and teamwork to get through it all from start to finish. My squad, which was integrated, made it through in about six. We were about ten minutes behind the pace of the squad that was the fastest in our group. There was one obstacle which is a monkey crawl on a rope that crosses a river. When I got across, the instructors were all yelling at each other and I couldn't understand what they were saying. I was worried that someone was dying or that I had done something terribly wrong. When I asked, though, they told me that I was only the second female *ever* to make it over that obstacle. Of course, the next two females fell. [Laughs] We had harnesses, though, and a big part of it is just trusting the fact that you're secure and you do, in fact, possess the physical and mental strength to make it across successfully. As soon as you start to doubt any of those elements, you're screwed! You've told yourself you can't do it, and you've lost the battle.

Another obstacle was a clay path created by water runoff. You had to go down this path into, literally, a stream of mud which was waist deep. Down this path, we were required to carry a casualty on a litter made of our own gear. Remember, there was nothing on this course to help us. We had to use what we carried and what we could create from the natural environment. So we carried a person on a litter down this incline, across this river of mud and then up another steep incline of running watery mud. When we started out, I asked to lead with the litter. Some of the squad said, "No, SSgt Rogers," but I said, "Let me have it now while I have the energy. Let me carry this and you big guys can save your strength for the next task which might be more physically demanding." So, we did with one of the less strong females on the back end. It was awesome even though, going down, I really carried the bulk of the weight while the girl in the rear primarily balanced it. I never really thought about it until I was done. I had made what I thought was a logical plan and I stuck to it. We, therefore, as a team elected to use certain strengths of our squad as the situation demanded it. This is what Marines do. It doesn't matter if you're male or female. It's about dealing with the resources you have at hand.

We did another scenario or two and then we got hosed down, literally, and were told to go back to our tents. We were given MREs and I went back to my tent. I looked around and there was nobody else in my tent. I was like, "What's going on?" So, I went to the males' tent and they asked me what was wrong. I said, "There's nobody over there." One of the sergeants said, "Oh, yeah, so-and-so hit her head, and so-and-so broke her arm, and so-and-so got hypothermia, etc." I was the only female that made it through that course. It was crazy! Now, there were plenty of males that suffered similar injuries and all, but the fact that there were only eight females that started the course really made the situation look odd when I was the only one left standing in the bivouac. The guys didn't really care. They were happy I was still there with them, but nobody made a big deal out of it.

Captain Kami also recognizes the importance of physical competency, but has never felt the need to "prove her manhood" beginning with her TBS experience:

At TBS, the issue of gender and sexuality was definitely more pronounced. I don't consider myself a "manly girl," not even close, but there were certainly the "girly girls" that were already starting to play on the male's interest in them sexually to get attention or favors. These were, of course, the girls that did the very same things I did at OCS. The only difference was that, now, the games men and women play with each other were added to the equation. My attitude was that if a guy wanted to shoot the SAW, he could carry it for the 15 miles. I had no interest or need to shoot it, so I was content to carry a lighter weapon. I would never relinquish a weapon I was assigned, but if a guy wanted to choose it, he could have it. Oh, yeah, and *clean* it, too. [Laughs]

All but a handful of female Marines describe the difficulty they experience in physically keeping up with the males after initial training. Most express the belief that if a female tries her best, doesn't quit, and can perform well in her job and on her PFT, the issue of physical weakness is not over-inflated into a compromised leadership issue. In sum, for most, physicality isn't a deal-killer but remains a huge part of her overall acceptance by the larger group. The key is her commitment, which manifests through kinship and leads to equivalency with her peers.

The Challenges & Benefits of Separateness

Recognizing the separateness of females from each other within the Marine Corps, SSgt Manda relates her experiences related to both issues of being totally alone, separate from the males as well as the only other female in the unit, while at Engineer Equipment Operator MOS school at Ft. Leonard Wood, Missouri:

One woman had graduated the week before I arrived and, as a female, I was assigned to live in the upper deck squadbay over the detachment office *by myself* for weeks before another female arrived. It was March. The weather was cold. The building was old and creaky. It was horrible. I'm not scared of ghosts or anything, but there was definitely a weirdness going on for the five weeks I spent living alone in this huge, empty building by myself...The only other female Marine on base at that time was a female Staff Sergeant, an admin MOS, with these big-assed fake nails and perfect skin and hair. This was the mid-80s, okay, *before* perfect nails were the "thing," but she had perfect nails, perfect hair, perfect make-up. So, anyway, here she is: my Woman Marine "role model." What a pathetic joke! I came back with all those filthy guys every evening after classes, exhausted, covered in mud, suffering from no sleep, of course, 'cause I still had to stand my watches like the guys. Was she out in the snow or mud at 5:30 in the morning for roll call? Or 6:00 a.m. for PT? No fucking way! What's *she* gonna tell me about life in the Corps? That experience really put a bad taste in my mouth. The job itself was actually pretty fun and I was great at it, but I felt alone from that point on and it never changed until I changed my MOS at my re-enlistment. I had, after all, come in open contract like a dumb ass, and in spite of it all, I loved the Marine Corps and wanted to give the experience another shot. That was when the Marine Corps experience finally began to feel more like a "life" than some kind of freakshow I was in.

In isolation and in the absence of a credible female role model over her first four-year enlistment, Manda's Marine Corps experience only improved after she changed her MOS from engineer equipment operator to linguistics. Her feelings of isolation and disgust with the only other female she encountered did not discourage her away from the Marine Corps experience as a whole. Other females, however, express different experiences and choices.

MOS Proficiency, Satisfaction, & Reputation

Unlike SSgt Manda's situation, which represented a combination of undesirable circumstances, Corporal Vanessa describes her current MOS and its specific downsides:

I am a 6114, helicopter mechanic. I did want fixed wings, but the job is really cool. Where else would I be able to do this? Being a Marine has enabled me to do this and I love it! My proficiency, though, isn't so hot. After a year, I am still not at a level where I need to be; I'm just not grasping it. Within the shop, it's split in two. There is the maintenance side, which is the mechanics, and the inspection side. If you're a good mechanic, they're not going to make you a plane captain, which is an inspector. They're reluctant to move anyone from the MOS if the Marine can be a contributing member of the squadron. I'm a weak mechanic, but I am a great inspector…There are about 15 females, total, in the squadron of about 370 Marines. There aren't many of us. Generally, women have a very hard time in the flightline shops. Sometimes I think it's a lack of real mechanic skills or maybe just the desire to perform those required skills and perform them well. On paper, I think, a lot of us score well on the mechanical part of the ASVAB, but when it comes to actually living day-to-day in the MOS, a lot of women just don't do well, and I've seen this in many other shops. We definitely have males in the shop that aren't as good in the MOS as they should be but, with the females, it's so much more obvious because there are so few of us. The females get categorized and that carries over into the rest of the experience and even to all other females.

In addition to physical and social isolation, CWO3 Leona describes her MOS, aviation ordnance, as a challenging job that requires physical and mental toughness:

> When I got to my first duty station, I saw the bias right away. I have always been an aviation ordnance Marine. Being aviation ordnance, it's so male-dominated. I know that sounds ridiculous to say because the whole Marine Corps is male-dominated, but it's even more obvious as a "mixed gender" MOS because there are so few females, and they usually get weeded out during their first enlistment. The fact is that it's a very physically demanding job with heavy lifting and long hours. It's a very boy's club-type MOS with a big emphasis on male camaraderie, which is great because it's a very tight-knit MOS and very loyal. But, if you can't hang, like you're not just as much of a smart ass as they are or if you can't take it as much as you can dish it back in their face or if you're thin-skinned, you're gone. You will not last long in that environment. It's a great community, but it takes toughness, mentally and physically, and a certain degree of callousness to make it work.

Subtle Displays of Power

SSgt Tomasa describes her position of chow hall power over the "superstars" of the Marine Corps, the awe-inspiring Marines of 8th & I:

> So, I checked into 8th & I, which is almost all grunts who are there as body bearers, the Silent Drill Team, and the Band. The only room they had available was on the floor with the body bearers, so they stuck me in there. And, all throughout the night, I'd get knocks on my door. When I opened it, they'd be like, "Yeah, there's a *female* in there." I was amazed that they were so shocked that a female was on their deck. All that week, they were stunned and confused. Maybe some were angry, I dunno. But, by the end of the week, I'd get a knock and they'd say, "Hey, M. we're going to go get some food. Wanna come with?" There were other females, administrators and cooks, attached to 8th and I, but they all lived out in town. I was the only one actually living with these guys.
>
> The body bearers were *so* funny. They treated me like their little

sister. I guess it's always been that when they came to the mess hall, they got to eat whatever they wanted. If they wanted a 10-egg omelette, they got it. But, they'd come to me now and say, "I want an omelette with six eggs." I'd be, like, "No. You only get two." The other cooks were whispering, "Damn, girl, *give* it to him. You've got to give them what they want." I was like, "No. The order states that we'll give them two-egg omlettes." If they want more, they can come back and get more. They're all big, *huge* guys, but I'd tell them, "Go eat what you got and come back if you want more." I knew these guys and they all knew me, but before we got cool with each other they got their two eggs and came back if they wanted more. After that, though, I hooked 'em up with whatever they wanted, but they had to understand who we were to each other.

They were all towering over me, all muscles and looking *fine* just like the recruiting posters. They even had their own *groupies*! They were local celebrities. It was my goal, my mission, to bring them back down to reality. So, when they came on mess duty, I treated them like I'd treat any other Marine at any other base. I even had a marcher come up to me later and say, "I remember you. When I was on mess duty and I pissed you off, you'd make up cut onions 'til I couldn't cry anymore and I stunk for days." There were marchers and the Silent Drill Team. Everyone saw them as these awesome celebrities, but I just saw them as my little snot-nosed kids coming into my chow hall to eat. I loved to get them on mess duty 'cause they came back down to reality for awhile. We were all fighting the same fight, just with different roles.

Even Cpl Dakota, a self-described "lowly" admin clerk, describes her sense of empowerment and opportunity to exact a playful revenge on the establishment:

I had gone to college briefly and majored in English Literature. At Lejeune, one of my many responsibilities was writing awards like Certificates of Commendation and stuff that were read at the Friday morning Company formations. I really loved that part of my job because I would use alliteration and create tongue twisters in them so the person reading them might get tripped up on the language and sometimes even get flustered about it. There was nothing *wrong* with the citations, of course. They were beautifully written, if I do say so myself, but when someone else, like the Admin Chief or the CO was reading them on paper to sign them, they didn't get the full effect of how the words would actually *sound* being read out loud to the formation. [Laughs] There was one darling little adjutant that I would get every time and he'd get all red in the face. He was damn cute to begin with and him becoming all red-faced and stuttering just made him absolutely *adorable*. I had to spread it out, obviously, but after the first few times, Marines being how they are, caught on to what was going on. The XO and CO thought it was pretty funny, but they did ask me to go easy on the junior officers, and, of course, I did so, but grudgingly. It was great while it lasted, though. I hope the adjutants I messed with have fond memories of those Friday mornings reciting my alliteration before the company and sometimes getting tongue-tied. Those were the good old days.

LCpl Cheri illustrates her pleasure with her MOS as an Engineer Equipment Mechanic and the power she draws from her competency in it:

I came in as a 1341 and I love my job. It's something that I never would have done as a civilian, but I feel like I am a more competent person because of the skills I have learned. Before I joined, I knew nothing about mechanics. I *never* wanted to be the type of female that wanted to be stuck on the road and be totally clueless. I *never* wanted to be the type of female who'd show some leg or twirl my hair to get a knowledgeable man to take care of me and my problems. They respect me because of my personality. I have set the boundaries. The first is: don't assume you know me when you don't. The second is: don't assume that you need to take care of me because I'm a woman. I am a Marine and you will deal with me as such.

Ultimately, too, if she chooses, a female Marine *will* have her day even in the face of adversity such as a hot-headed male lieutenant. Cpl Michelle describes an incident involving a fellow female heavy equipment operator on Naha Pier in Okinawa, Japan:

One time we were on an operation at Naha Port in Okinawa and there was me and another female heavy equipment operator and a few other of our guys using 4,000-pound forklifts to offload the maritime pre-positioning ships, rotate the equipment, and then reload them. The weather was beautiful. It was spring, the sky was blue, and the op was going as smoothly as any I'd ever been on. It was really a pleasure to be working at that pier, that time, with a great crew. It was one of those beautiful times that you remember all your life. Then, the next morning, a butter bar [Second Lieutenant] showed up, new to the rock [Okinawa], too, of course, and started changing around a pattern we'd been running with the gear for about a week. At first, we thought he was just another box kicker [Supply MOS], but we'd been working with them the whole time already and everyone was fine with the progress we'd made and how we were moving their gear around. Anyway, this idiot starts getting in all the operators' shit, telling us we were too slow, that we had no idea how to stage his gear properly, etc. That happened for one entire day, and it was just too damn beautiful to have an asshole like him try to bring us down. That evening we found out from one of the longshoremen crews that he had something to do with aviation radar. The next morning our gunny said to just ignore him, that his stuff would be off today and he'd be gone soon enough. See, sometimes, we didn't know what was in the boxes we were hauling, but it did make sense that he showed up when we were pulling that type of stuff out of the ship's belly. So, sure enough, this guy shows back up on the pier, flapping his gums all over the place about our incompetence, marching up and down the pier in his brand-new hardhat and Mickey Mouse ears like he was somebody. Well, about mid-morning, Sergeant Fasthorse was hauling another set of boxes down the ramp and this guy started running toward her and waving his arms for her to stop. He's an officer so, as soon as she saw him, of course, she stopped and dropped one of her ears so she could hear what he's saying. I was waiting to go back into the ship, and out of the corner of my eye, I see our SNCOIC heading toward both of them and I'm thinking, "Oh, boy! There's gonna be a fight now." Anyway, the lieutenant is up in Fasthorse's face, yelling at her at point-blank range and she's yelling right back at him. Everybody on that pier including American civilians and the Okinawans that worked there started gathering around, all big-eyed, staring at this insanity. [Laughs] It was absolutely crazy! Well, before Gunny could even reach them, Fasthorse puts her ears back on and motions for him to get the hell off her lift and she starts backing up with her load with the lieutenant hanging onto the side of the lift, calling her a stupid bitch. Oh, we all heard that little comment. So, he's now obviously trying to take the wheel away from her and she slams on the brakes. Although she was just creeping along, she was backing down a ramp on a ship and those ramps can be pretty steep. When she stopped, I'm sure for safety reasons as the asshole was grabbing at the wheel, the top box on her load slid just enough to the left side that its weight shifted enough to carry it completely off the other boxes. It bounced once on the ramp, split open like a piñata, and then the entire contents dumped into that beautiful clear blue water! Watching it was like slow-motion. I will never in my life forget what happened next. In the total silence... well, not really because it's the biggest port in Okinawa, but it seemed like it...Sgt Fasthorse, her Cherokee all coming unglued, looks at the lieutenant and says, "Nice fucking job, there, sir," and she just cracks up laughing. The lieutenant screams at her, "YOU FUCKING CUNT!" And, she laughs harder in his face. Thank GOD, Gunny was there by that time! He's a pretty big guy and he pulled the lieutenant away from Fasthorse's forklift and was trying to speak rationally to him, but the butter bar kept pulling away from him, trying to get back at Fasthorse. He was in a total meltdown screaming about "three million dollars worth of equipment" and "dumb bitch will pay for this" and I don't even remember what all else. He threw his brand-new helmet on the ground. He threw down his ears. By that time, though, Sgt Fasthorse had calmly scooted her lift down the ramp with the other boxes and proceeded to stage the stack on the pier like we'd been doing for days. Always the prudent professional, though, she then parked her lift, took off her hearing protection, and walked coolly over to the gunny and the insane officer to see what kind of trouble she was probably in. As she did, though, all of our guys and me got off our lifts and went with her. Two MPs appeared and collected the lieutenant, immediately leading him toward the port headquarters, and asked the gunny and Sgt Fasthorse to come with them. Before they did, though, Gunny said, "Fasthorse and I got this covered," and she's looking at him, like, "Yeah, okay, it's a done deal." And to us, Gunny says, "Y'all get the ship done." He grinned then. Relieved, we, the operators, responded, "Aye-aye, Gunny," but we waited a bit to make sure no reinforcement was required. Gunny had that mischievous look in his eyes, not unlike Sgt Fasthorse, and it was beautiful. The story became legend on the island. Although there were never any charges filed by anyone or an inquiry into who was right or wrong, there was always speculation and rumor that Fasthorse dumped that shit into the port just to piss off that obnoxious boot lieutenant. Suffice it to say that she ruled the shop from then on. Always known as a competent operator and a damn fine Marine, her authority had ratcheted up a few more notches. Nobody ever questioned or fucked with Fasthorse. As a female myself, a fellow HE, living in the barracks, later, when I asked her if she did it on purpose or not, she just smiled and said,

"If I had really wanted to piss him off, he would have ended up in the bay with his damn radar shit." [Laughs]

USMC: The Ultimate On the Job Training

These stories exhibit the various coping strategies employed by female Marines as they negotiate their personal and professional lives in the sex-skewed and operationally demanding environment of the United States Marine Corps. While learning and enacting their occupations on a daily basis, females manage various aspects of their very being: physical, personal, social, and occupational/ professional. In all of these situations, it is fundamentally her sex, Kanter's "token," that is disruptive and often defining at this early stage in her career.[22] Caught in conflicting role demands,[23] a young female Marine must develop coping strategies to mitigate the impact of her sex/gender. This enables her to be wholly competent, secure in herself and her relative position, and to ensure respect and trust by her male and female colleagues. Relationships forged between equivalents in this military environment are genuine, rewarding, and lasting. A sense of kinship and concentration on individuals' contributions, unique talents, and potentials unrelated to their sex, fosters a greater sense of equity and cooperation.[24] In survival situations and warmaking, as in life in general, males and females are equally important to the mission.[25]

Endnotes

1   Schwartz and Rago, "Beyond Tokenism," 71.

2   A reference equating the transition from initial training (death of the self) to the Marine Corps (attainment of glory) made from the *Catechism of the Catholic Church*, (New York: Doubleday, 1994), 291.

3   Williams, *Gender Differences*, 60.

4   Emerald M. Archer, "The Power of Gendered Stereotypes in the US Marine Corps," *Armed Forces & Society* DOI: 10.1177/0095327X12446924 (May 2012): 22.

5   Schwartz and Rago, "Beyond Tokenism," 72.

6   See David M. Buss, *The Evolution of Desire*, 145-147, where he explains the vastly differences in perception and interpretation of sexual intent by men and women.

7   Ibid., 80.

8   Kanter, "Effects of Proportions," 976.

9   Buss, *Evolution of Desire*, 26.

10 Kim M. Elsesser, "Does Gender Bias Against Female Leaders Persist? Quantitative and Qualitative Data from a Large-Scale Survey," *Human Relations* doi: 10.1177 /0018726711424323 (November 2011).

11 Leah D. Sheppard and Karl Aquino, "Sisters at Arms: A Theory of Female Same-Sex Conflict and Its Problematization in Organizations," *Journal of Management* DOI: 10.1177 /0149206314539348 (June 2014).

12 Buss, *Evolution of Desire*, 214.

13 Ibid., 186.

14 Ibid., 80.

15 Ibid., 95.

16 Brownson, "Battle for Equivalency," 6.

17 Meštrović, *Trails of Abu Ghraib*, 25.

18 Sarah Jones, "Sen. Gillibrand Takes on the Military Rape Culture and Gets Told She's Not an Expert," *Politicusa*, November 17, 2013, http://www.politicususa.com/2013/11/17/senator-kirsten-gillibrand-takes-aim-military-rape-culture-told-expert.html.

19 "RAINN Urges White House Task Force to Overhaul Colleges' Treatment of Rape," Rape, Abuse &

Incest National Network (RAINN), NewsRoom, accesed May 15, 2015: https://rainn.org/news-room/rainn-urges-white-house-task-force-to-overhaul-colleges-treatment-of-rape.

20 Stremlow, *History of the Women Marines*.

21 Brownson, "Battle for Equivalency," 9.

22 Janice D. Yoder, "Rethinking Tokenism: Looking Beyond Numbers," *Gender and Society* 5 no. 2 (June 1991): 189.

23 Schwartz and Rago, "Beyond Tokenism," 71.

24 Ibid., 76.

25 Dowling, *Frailty Myth*, xxxiii.

Chapter Five

Worlds Colliding: Living in Separate Spheres as Women & Warriors

Once they become United States Marines and have negotiated the initial trials of TBS, MCT, and MOS schools, female Marines must navigate the even more precarious terrain of their relationships with civilian women, men, and the larger society.[1] They seek inclusion by the organization via the concepts of kinship and equivalency.[2] Through this process, they develop an awareness of how their presence in the Marine Corps is often criticized and manipulated by individuals within and outside the military environment.[3] For example, they struggle against the male Marines' wives' mistrust and hostility.[4] If they had not realized it up to this point, savvy female Marines begin to understand that they exist between worlds, such as: the military sponsor-dependent, the civilian-military personnel, and male-female actors within and interacting with the organization.

In this isolated yet interactive environment, female Marines embody alien representations to those with whom they interact. Kanter asserts that Simmelian assumptions determine processes and narrow the universe of interaction possibilities of tokens, which determines the interaction dynamics between tokens and dominants. She is correct that tokens in the Marine Corps embody higher visibility as well as polarization and exaggeration of differences between female Marines and male Marines as well as the civilian community.[5] The data confirm that female Marines occupy a position of higher visibility and that gender-specific polarization does, in fact, exist. Ultimately, however, female Marines' experiences in the modern military and, specifically, in warfighting negate differences between the sexes in the Marine Corps while compounding negative interactions with civilians. Through the contemporary experience of warfighting, female Marines match, equivalently, the experience of their male counterparts in terms of commitment, sacrifice, and empathy for their fellows as well as the enemy. The kinship bonds among Marines, male and female, solidify insider/outsider affiliations.

Marine Females' Intersection with Civilian Females

Rather than ambivalent, many female Marines express passionate feelings about their achieved status, their command of their lives, and their relationship to those on the "outside" of their experience. SSgt Tomasa, for example, expresses highly charged opinions about civilian females and their presumed "role" in the Marine Corps:

> Female Marines are simply awesome. What I *fucking hate* is those shirts and bumper stickers that say, "Marine Wife: The Hardest Job in the Marine Cops." That shit makes me livid and I throw up in my mouth a little! *Damn*, I hate that. I need to make up my own T-shirt that says, "The toughest job in the Marine Corps is being a Marine married to a Marine with kids and going to school." Beat that! It's like everyone forgets about us sometimes. Here's my life, if you can really call it that: I wake up and I get my kids ready and off to school. I'm off to PT and then work, or directly to work. I work all day as a cook. At the end of the day, I may have a LCpl come to me and tell me his wife in East BFE, Missouri, or wherever is getting evicted and they have no money. So, I get working on solving that problem to take care of him. I get that underway and leave to pick up my kids. Drop the kids off at the babysitter and I'm off to school. Come home from school, picking up the kids on the way. I get home, and we do baths and get ready for bed. This is *assuming* that any homework was done at the sitter's which we then check and correct as needed. I get them to bed and then I get to do the dishes or laundry or whatever domestic tasks need to be done. Then, I get to have some free time to study or prepare for the next day. If I'm lucky, my husband will be around to help with the child-shuffling and the babysitting issues. But, as a Marine also, he's often not available to pick up the slack. And, throw into this mix the rifle and pistol range, the field, field days, deployments, schools, inspections, and the myriad other time and energy intensive aspects of a Marine's life. It can get pretty hectic.

So, sorry to be so blunt, sugar, but being a Marine's wife is *not* the hardest job in the Marine Corps!

SSgt Robin also describes her impatience and frustration with civilian wives:

These women around here [Camp Lejuene area] just want their husbands back. I'm not saying that's wrong, but it's amazing that that's all they really have in their lives. Like, they'll say, "Oh, your husband's deployed, too? Oh, that's so sad. What do you do, sit at home all day?" "Um, actually, no. I'm a Marine, too. I really have better things to do. I'm SSgt Vale." It's hilarious. The key wife network, because I'm on their "list" [as a spouse], will call me and ask, "Are you okay?" I tell them, "Yeah. I'm fine. Have you called Lance Corporal So-an-So's mother to ask if she needs anything?" They think I'm a bitch and I don't care. For my husband's unit, I'm a volunteer, too. I call all of the single Marines' parents and give them the updates. They're the ones forgotten in this mix. They're the ones sitting at home, watching CNN and talking about how "good" or "bad" the Marine Corps is because they have or haven't heard from their son or daughter. The key wives network doesn't really give a shit unless a Marine has a wife, and most of them are sitting here in Jacksonville getting their nails done at a sweatshop where all the scuttlebutt is about the deployments. They're so limited, but they think they're really sacrificing. It's sickening.

GySgt Claire asserts that her place, her home, is in the Marine Corps, which is a problem for female civilians because they can't fathom her experience:

The most important thing I would like to say is that civilians look at females as their sisters and daughters. They don't look at us as individuals doing this job. They lump all females together into this "female" category and they compare that to the "man" category when they decide who can or should do what. They need to realize that women can do everything men can do whether society is ready for that or not. The mentality of female Marines is a lot stronger, I think, than that of other females. We're more flexible in our thinking and we multitask more effectively. We get the job done and we get it done right. When they picture a Marine, all they picture in their minds is this beautiful male storming into combat. When they think of a female Marine, there is no picture.

Are we firing heavy weapons? Are we carrying our children? Are we guarding a convoy or detonating a mine? Yep. We're doing *all* of those things and more every day. We're also shopping and doing the laundry and having our nails done. We have created a strong presence in the Marine Corps, and I am looking forward to that role expanding further. I am a lady, but I am also a competent Marine and no one can take either of those things away from me.

SgtMaj Brittany describes a situation in which she went head-to-head with her Battalion Commanding Officer's wife:

It seems to me that a man will only listen to two women: his mother and his wife, if he's married. I think he's just not going to listen to another woman. For example, I am a Battalion Sergeant Major [highest ranking enlisted Marine], the senior enlisted advisor, to a Lieutenant Colonel, the Battalion Commanding Officer, but in so many cases, my advice was discounted, and he would listen to his wife. It was very obvious that his whole demeanor changed when she came around. It took me at least two years to figure this out and she was not prior service, so I don't know how he thought she was "smarter" at this game that I am. I don't know, but on more than one occasion, I've had to do damage control after these incidents. It's just not right.

Captain Kami expresses outrage by the assumptions that the Marine Corps seems to impose upon her:

Every time I join a new command, I am given information about joining the Officers' Wives Club. I am appalled by this because I am *not* an officer's wife, I *am* a Marine Corps officer! Even if I *was* an officer's wife, I would *never* join that organization because my status as a Marine officer *myself* would, obviously, override my status as "wife." *Every* command I have joined has invited me to join the Officers' Wives Club. Every. Single. One. I can't relate to those women as an officer's wife because I'm not one. I can't relate to them because I have children, which I don't. I don't know if this is just some standard pamphlet that's in the "Welcome Aboard" package or what. I do know that, as a Marine, I don't have time to do officers' wives' "things." I have real, Marine Corps "things" to do. It's ridiculous and it's an absolute insult to me! I usually say something when that's put to me, someone assuming that I am an officer's wife, but I don't hold a grudge and I don't get all inflamed about it. The sad thing is that they don't even realize that they have insulted me. It's really just sad, but yet another thing we/I go through in this hard-to-define status as a female Marine officer.

CWO3 Phoebe identifies the ethic, the *esprit de corps*, and civilian jealousy as the divisive factor between civilian women and female Marines:

What's incredible is that when you look at us you see women but, more importantly, you see women of power. I think if someone wants to see the most beautiful females in America, go to 4th Battalion [MCRD Parris Island, SC]. Those are the most together, most physically and mentally strong, most sincere females you'll probably find. You need more? Go to Iraq and talk to our Marine Lionesses there. We're mothers. We're wives. We're sisters to each other and to our male Marine brothers. We're daughters to those incredible men who helped raise us to not be afraid to take that step outside the box of feminine expectations and to seek more for ourselves…In my limited experience with people outside the Marine Corps while I've been in, I can honestly say that I have not met the caliber of person, especially not the caliber of female, that I have encountered in the Marine Corps. We're a unique group. Other women, say women that live in a given suburb, have kids that go to school together, or they were cheerleaders in high school or whatever. They'll never have the common bond that female Marines share with each other and with male Marines. In 20 years, I have never been disrespected by a male Marine. Civilian females, though, have looked at me like I'm some kind of nasty bug they'd like to squish under their stiletto heel or Birkenstock sandal. I think they hate me because they're jealous. They took the "easy route" while I was breaking barriers. I really don't care about what they think, though. I live every day really happy with my choices. I am so proud to wear this title and it's all about *me*. That's what I love so much. I have made my own way and I am so proud of that.

Due in large part to the uniqueness of their occupational choices and experiences, many female Marines, as discussed previously, describe a sense of isolation as well distrust and animosity projected by female civilians toward them. As GySgt Claire previously stated, civilians simply do not possess a "picture" of female Marines; they are an enigma.

Civilian Mistrust of Female Marines in the Work Sphere

In the military work sphere, female Marines develop a high degree of trust with their male counterparts as well as interaction norms that distinguish between those who are similar and equivalent, Marines, and those who are different and unequal, civilians and "women."[6] The issue of female Marines' sexuality, of course, becomes an issue as described by MSgt Madeline:

> At my first duty station, there was still that mind-set amongst the wives that I wasn't a "real" Marine. A female Marine in the office was a distraction. She was there to sleep around, to distract their husbands[7]…It's always been tough with me being in an office with a bunch of men. The wives saw me as a threat. It's changed since then, I think, because wives and girlfriends realize that I'm not there stalking their mate. I am there because the Marine Corps has assigned me to be there. I'm married and have kids, so I don't see that "look" anymore. Maybe, too, as I've gotten older, I've just learned to deal with it better. I try to make an effort to get involved with my Marines and their families so they can see my role as a supervisor of Marines, to see that I am there to assist *all* of them.

> If we get a young, pretty, flirty female Marine in the S-1, it does create concern. The males obviously take notice and behave differently than before she arrived. The wives, girlfriends, etc. get suspicious and I have to exert my authority over the situation to ensure that it doesn't get out of hand. Now, as a female Marine myself, I should never *assume* that a female Marine would come into that situation with motives to be disruptive. However, it *is* my job to ensure that good order and discipline remain in place, so I assume the worst and hope for the best, making corrections to both males *and* females as appropriate along the way. In fact, it's probably better that I am a woman because I probably know every little game she might be inclined to play, and can shut it down in a heartbeat. She needs to know that when you're in uniform and especially in my S-1, it's all about business. You're a Marine, nothing different, nothing less.

Whether completely incomprehensible by civilian females or construed as little more than paid vixens to seduce "their" male Marines, female Marines experience a range of personal and professional challenges in the workplace from external sources. Female Marines step beyond the everyday expectations of sex and gender relations, embracing a truly enigmatic lifestyle, a situation that cultivates mistrust and contempt by civilian females, male and female alike.[8]

Kingsley Browne asserts that women's greater egalitarian tendencies may make them resentful of a hierarchy that automatically places some individuals above others.[9] This research supports this assertion. A wedge remains in place between the two groups, those females who *live* the Marine Corps lifestyle and those women who live *vicariously* through their male Marines as a result of personal deprivation.[10] Desirable mates are always in short supply. Glamorous, interesting, attractive, socially skilled people are heavily courted and rapidly removed from the mating pool.[11] From the previous chapters and the existing literature, it is evident that Marines, both male and female, possess characteristics as highly desirable mates.[12] Mate poaching would certainly be a concern for civilian females.[13]

Female Marines' Concerns about Civilian Control of the Military

The U.S. military is not a perfect institution comprised of perfect individuals; however, it strives to achieve excellence in military leadership.[14] Military sociologist Charles Moskos asserted after the 35th Annual Tailhook Association Symposium debacle in 1991 and the Aberdeen Proving Ground sex scandals in 1996 that it had become an article of faith for feminist spokespersons to hold that sexual harassment would come under control only when women were no longer regarded a [*sic*] second-class members of the military–that is, no longer excluded from the combat arms.[15] Unless the service branches articulate why they should not, on January 1, 2016, American women will enter the combat arms of the United States military, a notable departure in world historical terms.[16] Interestingly, "final approval of human-research requirements"[17] must be met before females can officially participate in a "simulation of an operational environment," indicative of civilian control over the military.[18] These females, obviously, must prove themselves capable yet again.

Many of the female Marines interviewed whose stories are told here experienced combat in Iran and Afghanistan in the past decade.

They remain wholly ambivalent regarding the reciprocity of combat inclusion to eliminate sexual harassment, assault, and rape, believing the issues of combat inclusion and sexual assault to be unrelated. They further do not believe that being sanctioned combatants makes them "full citizens." G.L.A. Harris' asserts that our orderly, self-contained politically defined system should rule the day, stating that "if we pride ourselves in being the model and steward of democracy…we must be the first to live it, honor it, cherish it, and do all that we can to protect it." She further asserts that the "journey toward equality, full citizenship, and hence full agency for military women and correspondingly for American women in general will no doubt be rife with unforeseen trials, much have largely been defined the arduous road already traveled…the revolution must continue."[19] In stark contrast, my research asserts that female Marines do not give a damn about being a "model or steward of democracy" on a symbolic scale and they are not engaged in a revolution. They simply have a job to do. SSgt Nadine summarizes this astutely:

> I love being a female Marine. It's not like being a police officer, where they protect the people of their town or their community. As Marines, our job is to protect the whole United States and other people around the world. We do what the president tells us to do. It is exciting to be in this position, knowing that we have the ability and the responsibility to take care of the United States and the world. We knock down the bad guys when we have to, and we build up the ones that need help. It's always been this way, and I think it's still a really important job and I'm honored to be a part of it.

Without exception, female Marines in this study adamantly express anger toward outspoken feminists, politicians, evangelical preachers, and organizations such as DACOWITS who, in their opinions, do not express their preferences on many issues related to the U.S. Marine Corps and their service within it.

Feminists, Legislators, the Media, & Female Marines

SgtMaj Brittany speaks openly about her disappointment with "outsiders" dictating policy to an organization that they do not understand:

Our government, our legislators, for all the wonderful things they, most of them, do for us have never walked in our shoes. I don't think there's a single one of our female U.S. representatives who's ever been in the service and, as a result, I don't think they have the proper ideas of what it's really like. I've heard of other lawmakers trying to fight for women to get in the cockpit, and that's fine. I think DACOWITS is a good thing, but I don't think it's utilized properly. I think it's putting lipstick on a pig. They go in there to "investigate" and hear what the military wants them to hear, but it doesn't do anything. What they hear from PAO or the high-ups is not necessarily the reality, and you can't make informed judgments with inadequate information. I think people who have never been in my shoes shouldn't try to speak for me. Feminists, you know, when it comes to the hard core-ness, like "Get outta my way because I'm a woman," I think they do more damage, more harm than good. Before you can get out there and say you're for the rights of women in the military, sister, you need to have worn my combat boots. That bothers me. If you don't know what it's like on the inside, how do you know what the inside needs and wants?

SSgt Lynette, a Motor Transport Marine with ten years time in service, agrees:

I think a lot of the politicians are rich and they can't understand any other way. They went to college instead of the military. Their kids went to college instead of the military. They look down on those of us who enlisted as a way to make our lives better since we *didn't* have the financial resources that they have. I think a lot of politicians and academics view the military as a dumping ground of sorts for "those poor kids with no other options." From what I've seen and experienced, though, we don't want their pity. Especially after that first enlistment, we're here because we know what we've gotten ourselves into and we're staying because it's what we've come to love. I've heard civilians say, "Oh, he or she only stays in the military because he/she can't do anything else." I say, "Come on, then, and walk a few miles in my combat boots," and we'll see who can and can't get by in one world or another. They speak from ignorance, the politicians, the academics, and the other civilians who say these kinds of things. They just can't understand it because you can't really "get it" until you live it. Until you know the kind of homesickness, the kind of hunger, the kind of thirst, the sense of responsibility that comes from being a

Marine, you'll just never "get it." No amount of liberal arts studies or committees or lobbying will ever change that fact.

SSgt Robin admonishes the media and calls on Americans to" wake up":

It's been awhile now, but in 2003, it was horrible, the whole anti-war thing. In Yuma, the news media came up to me and asked me, "How do you feel about people not supporting the war in Iraq?" I responded, "I would tell them, 'You're welcome.'" They were amazed, like I was supposed to be all teary-eyed and torn about what's going on. No! I said, "Don't worry about it. I'm going over there to fight for the right for people in this country to say what they want." You think the Taliban wants people to be "free" and "democratic"? No. They want to *kill* all of us because we won't adopt their religion and lifestyle. They don't give a shit about freedom of speech. If you're female, they think you need to be treated like any other breeding animal in a man's herd. Why Americans don't understand this, I cannot comprehend. Islamic terrorists don't want to "get along" in the larger world; they want you dead or subjugate you until you surrender to their views. This is *not* a "Barney and Friends" world, regardless of how much we all might want it to be.

Captain Maya offers a slightly different take on civilian/feminist advocacy and recognizes military equivalency without the requirement of a female to serve in the combat arms:

It seems like the women that are interested in "fighting for us" have their own personal agenda and it's not what we, as Marines, want. I mean, we're not fighting to get into combat arms MOSs. I have not met a single woman in the military that wants to be an infantry officer. No, thank you. But, that doesn't mean that you don't need me. I don't feel like I'm a second-class citizen because I'm not an infantry officer.

GySgt Mary Ellen describes her disgust with the military-civilian conversions currently taking place, concerned that Marines ultimately will come home from war to billets filled and out-sourced to civilians:

The Marine Corps is doing this whole military-civilian conversion and that was hard enough because the first actual *Marine* I encountered when checking into the air station was the Sergeant Major. The mail clerk is a civilian. The adjutant is a civilian. The guy who writes my fitness report is a fat, nasty civilian! He doesn't even know who I am. He provides the content for my fitness report. I work in an H&HS, and the Reporting Senior and the Reporting Officer for my fitrep [fitness report] are both civilians…because they are the S-4 chiefs, the supply people. Because we hand out bombs and rounds and stuff, we're considered supply. We're *not* supply, but because we "supply" things to people, they consider us "supply" and we go into this little pile of people over here. It's devastating. I thought this fat puke that's going to be writing my fitrep was the janitor! [Laughs] This is a trend going forward where DoD is trying to convert all of this area to civilian personnel. There will be no "military" at this rate.

All females who expressed an opinion, usually vehemently, about the involvement of feminists and other civilians who have never served, but are outspoken about females in the military and what they should and should not be doing, agree that the official policies should be made by those who can rightly speak to the issues, the females and males who have lived the experience. In particular, the recently invigorated concept of "women in combat" seems destined for implementation in 2016.

The Contentious Issue of Women in Combat

The voices expressed here offer opinions developed and enacted prior to, as well as, during the wars in Iraq and Afghanistan. Some of the female Marines interviewed were not even in Kindergarten when Operation Desert Shield/Storm initiated the subsequent cycles of the War on Terror. Still, LCpl Cheri describes her desire to deploy to war as soon as possible and why deploying is so important to her:

I will be in the war soon. I need to go because I joined the Marine Corps to serve my country. If I die a Marine, I know that I have done everything I can for my fellow Marines and for my country. I will go to Afghanistan because it's the right thing to do. I know a lot of people don't agree with the war, and to be quite honest, there are a lot of aspects of the war that I don't understand and I don't agree with. I was living in Brooklyn when those towers went down. Anyone who wants to look me in the face and tell me they know how that feels without *living* it really needs a reality check. I assumed responsibility for the reality when I arrived at Parris Island, and I am growing up here. This was my choice and I will do what is expected of me because I believe there is a greater good within what seems like madness. We are on the righteous side and ultimately we will prevail. Is it easy? No. Is it without cost? No. Is it worth it? Yes, it is, or I wouldn't be here. I'd be living the good life somewhere, watching "reality" TV, instead of watching the people I love deploy while my only action is to do my job to the best of my ability to keep the equipment operational and to pray that my fellow Marines come back alive.

Supporting why females should not be in combat, GySgt Iris identifies why she did not want to deploy to Iraq, making the distinction between carrying on a "name" [the USMC] and carrying on a "family":

Did I have any desire to go to war, to Iraq? No, 'cause I didn't want to leave my kids orphans. As a Marine, you don't have a choice. I don't know if the laws or rules or whatever have changed, but there was language for sole survivors or brothers not be sent together, but there is no restriction now against a mom and a dad both going to war. There is no "if one goes, one stays" directive. I don't understand that because it's like you're going to send both parents to war and they have no choice, what happens when the kids're orphaned? I mean, what happens to those kids? That just makes no sense to me.

Cpl Diandra, presently an engineer equipment mechanic and of a like mind with Iris, will be relinquishing her status in the Marine Corps at her EAS to be a mother:

In my brief career, I have received wise and respectful guidance, and I think it's because I try to be the best Marine I can be, at PT, on the range, or in my shop. My focus, though, is to be a wife and mother at the end of my contract. My NCOs and SNCOs are sad and they tell me so, but they also respect my decision…Yes, I was all gung ho, but being married and now having a small child has changed my mind about my priorities. There are a lot of women that can leave their children and their homes when the Marine Corps calls and I respect that, but I am not one of those women.

I don't want to leave my child to deploy to Afghanistan again. When I was single, I went and I appreciated the experience while there and what it means to me now that I am back. Being *in* the war has given me the greatest appreciation for my life that no other experience can. These are *hard* decisions we have to make! Mine is to be a mother to my children.

Other female Marines, however, remain more liberal in their thinking, focusing on the fact that females are still choosing to serve in the Middle East while keeping in mind that Marines of both sexes continue to die there, too. For example, combat veteran SSgt Tomasa, speaks to the reality of female Marines currently serving in combat zones:

Women are in combat now. If they want to be grunts, I think they should be allowed to do it. If a unit's getting mortared, no one stands up and says, "Okay! Time out! We need to get the women to safety!" There are people, male and female, who will be successful in those fields and others who will fail. Women are getting killed over there, too. The Marine Corps has to make the best assignments to, first, minimize the loss of life, and two, accomplishment of the mission. I put loss of life first only because you can't win if you're all dead.

Many of the female Marines interviewed expressed explicit concerns about the complete disregard of the laws of war by the enemy in contemporary military scenarios. Many view their primary roles as wife and mother. Whether their concerns are interpreted as fearful or purposeful, both valid and rational responses, this presents all the more reason to not send women *en masse* into combat scenarios as sanctioned offensive combatants. Similarly, if feminists and the broader society assume that *all* American females desire "equality" with males enacted through combat action, selective service, etc., they speak for a non-normative minority of military women as indicated by this qualitative sample. Conversely, as supported by these data, other factions that assert that females are only suited for sex- or gender-specific endeavors are equally wrong. Valor and self-sacrifice are not gendered.

Military Peer Leadership, Training, & Professionalism

John Allen Williams, political scientist and retired Navy captain asserts that a nation's political culture and shared memories condition both the military and its relation to the broader society. Military professionalism itself is shaped in a context provided by the states and its relationships with the outside world.[20] Radical feminists dictate that women can do anything men can do.[21] However, the changes in females' role in the military often reflect *de jure* rather than *de facto* equality, hence the importance of the concept of females' equivalency rather than the insistence on "equality" with males.[22] Laws cannot make anyone lay down his/her life for another; only kinship and the mutual respect, trust, and love it garners can.

As previously discussed, in the competitive military environment, female Marines experience subtle discriminations because of their sex/gender and how their sex/gender is perceived within the organization.[23] In some cases, it results from male benevolence, a form of mate protection in an extended kinship network.[24] At other times, it results simply from the expectation that a female should "know her place" in a warfighting organization. MGySgt Elizabeth's experience upon arriving in Okinawa as a PFC out of MOS school in the mid-80s reflects solidarity by necessity; a fundamental military task not taught to her was about to be tested:

> Thank God, they've made the changes they have from back in the day when recruit training for men and women was so different that the guys could realistically say, 'You're not as good as me because you were in boot camp for eight weeks putting on make-up instead of drilling with a rifle.' It was *true*! I got to Okinawa and had *never* drilled with a rifle before. A change of command ceremony was coming up in a few days. So, I went to the known-to-be-an-asshole Sergeant Major and respectfully said that I wasn't taught the Manual of Arms at Parris Island. He obviously wanted to make an example of me, the *only* woman in the platoon of engineers, so he wouldn't excuse me from the parade even though I had not been taught the drill. The guys liked me and were pissed. So, after work and all through the next two nights, they taught me what I needed to know. We knew it was about *us* as Marines, not about the lone deficient female. We practiced in our little squads in formation behind the line company's barracks. The guys seemed to

really enjoy teaching me, especially tricks to make the maneuvers easier for a physically weaker person. The day of the parade was *so* awesome! I nailed every movement like I'd been at 8th and I. The guys were very proud and we had so much fun celebrating at Okuma that weekend! It was one of the biggest little victories in my life. [Laughs] So, screw *you*, Sergeant Major! *That's* what the Corps is all about…Marines working *together* to make it happen, *not* playing your stupid little bullshit "girls can't play in my treehouse" games!

Female Marines' leadership is always initially questioned, even by themselves. SgtMaj Cadence describes the unique challenges of being a female in a man's world in the early 1980s:

Battling for my position was the most challenging thing I have faced in my career. In the heavy equipment field, I was always challenged and it was always assumed that I didn't know what I was doing, so there was that constant struggle to prove myself as a leader of Marines *and* as a heavy equipment operator. When I was at Landing Support Battalion, there were like nine staff sergeants, and I had two years time in grade. They didn't expect me to know anything, but I was senior to everyone else. So, I had to assert myself saying, "I *am* in charge of this. I am a SNCO and a competent operator and you *will* listen to me." Even the [male] supervisors at the time assumed that I wouldn't take charge of the shop and do what would have been expected of any male staff sergeant. I had to jockey for my position every time I went to a new shop because it was always, *always* assumed that I would defer to a male Marine even if he was junior to me.

CWO4 Renae also expresses frustration that her sex automatically negatively frames her leadership abilities:

If a female tries to say something against the status quo, you're seen as weak or a troublemaker. I've learned to hide a lot of things, especially my disgust and frustration at other Marines who are unwilling to step up and do the job, or my acknowledgement of something of concern. It's totally different for the male Marines, though. If they speak their minds, it's not indicative of their inability or reluctance to get along. Instead, it's them trying to "improve the system," so everyone listens. As a female, I don't have that luxury and I have to really decide what is worth putting out there and what isn't because what I say will be judged based upon my gender, first, regardless of the validity of my concerns or the way I express them.

Captain Kai provides a story of a fellow female officer failing on the rifle range and the complex issues of integrity and camaraderie:

There was one female in Iwakuni, Japan, with me who I didn't care for at all. She was overweight and weak, overall, so it was really hard for me to deal with her from the get-go. We were in the same company at TBS and in Iwakuni we were on the rifle range together. She was about to go unq, but the guys in the pits gave her a couple of extra points to give her a qualifying score. Instead of being grateful, she ratted them out and they got in trouble for lacking integrity. That was an interesting call when it comes down to Marine Corps leadership because they were really trying to help her out by giving her a couple of points, I think it was *exactly* four, to get her qualified. They were exhibiting loyalty which she did or, maybe, did not deserve but, yes, they lied on her score. When you have two leadership principles at odds, which one overrides? I don't know what the answer should be, but I'd be pleased that my fellow Marines cared enough about me to help me through a hard spot and be grateful for that while also telling them to please let me fall on my face the next time. Obviously, we don't want Marines cheating, but nobody died. Nobody was compromised other than the people involved, so it wasn't such a grievous thing to do, in my opinion. However, it blackballed her as disloyal and ungrateful, and potentially targeted all female Marines as "narcs," and got a lot of otherwise really good Marines in trouble. In addition to the half-assed officer she already was, she was labeled a *complete* shitbird from then on out. No one felt any loyalty to her again, and that's not a good position for a Marine Corps officer, male or female, to be in. She's out now, thank God!

GySgt Claire, a cook by MOS and the first female instructor at MCT, describes the prevalent expectation that females simply cannot lead and are destined to fail:

When I got to MCT, an extremely grunt-centric environment, one of the things they were most concerned about, was me allegating against them because they'd heard so many negative stories about female Marines. There were two of us who went to Hotel Company. This was 1997. When I got there, one of the things one of the males told me was, "Well, I didn't want to have anything to do with you women because I've heard what y'all do to male Marines." I asked him what he'd heard, and he said that we say that male Marines harass and rape us, etc. I told him, "Hate me for something that I did to you, personally, but until I show you that I'm not worth your respect, I expect a fair chance." So, we came in and we earned their respect. It was tense when we first got there, but the 1stSgt nipped it in the bud. He said, "Hey, they're here to stay and you better be able to deal with it or pack your seabag and ship." He was a good 1stSgt because when we got there, they went out of their way to get us up to speed on all the weapons. Once we learned it all and did it all, the guys were amazed and very respectful because we'd proven that we could do the same things they do. One of the concerns was that, when the females came in, would they be able to hang with the males? Our attitude was that we would make them hang because we're not going to accept them failing. It was me, a female corporal, and the two grunts who taught everything. My CO was really impressed. He told me in front of other Marines that I taught the best MARC19 class he'd ever seen in all of his years as a grunt. He said, "I can't believe you're a cook." We try so hard to prove ourselves that it's that much nicer when you receive that kind of accolade from someone who lives and leads in that lifestyle every day while you cook his chow. The guys were so impressed because they expected us to complain and whine and not want to do some things. Instead, we came in and accomplished the mission just like they did, and we expected the young females to do the same.

For female Marines, earning the respect of her male peers presents a constant challenge. Although male leadership deals with the same issues that female leaders deal with, the issue of a female Marine's sex/gender is always a concern. Questions of competency, authority, commitment, and trustworthiness automatically arise when a female attempts to actually lead her Marines as she has been trained and is expected to do.

Examples of Female Marines' Adaptation & Confidence in War

Female Marines, like their male counterparts, currently serve in every clime and place. Wherever they are, their challenges expand beyond the everyday military realm and into the personal/social. Caught between numerous spheres, SSgt Amber describes her experience from the perspective of a Marine Corps Public Affairs liaison, an American woman in war-torn Iraq, and the mother of five children:

In Iraq, I met amazing people. I also met incredibly stupid people. I met local people who were moms. I wrote an article while I was over there that made me not too popular. We had a son of one of the sheiks on the base and he had a little newspaper as one of his little pet projects, and he said, "Mr. Amber, would you write a story for my newspaper?" The men always use "Mr." as a sign of respect, so I wore it proudly regardless of the whole gender thing. So, we had lunches a couple of times a week and a flag officer asked me if I would come and take pictures of people. After awhile, they would speak to me. There was one female captain and me, but only men were allowed at the table. After a couple of the lunches, though, the sheik's son said, "Mr. Amber you will sit at the table." But, when I did, a few of the men got up and left. I apologized, but he shrugged and said, "Eat." Everyone in that country tried to make me eat. He asked me what I would write about. I thought about a recent convoy I had been on where we took water and food and other necessities to women and children in one of the towns not too far from the base. What struck me most was how moms do the same things there that we do here. What they wanted to talk about was their children. We all want our kids to be healthy and happy. So, I wrote a story about how similar Iraqi moms and American moms are. It [the article] had to be blessed by everyone in both cultures, of course, but I was pleased with it and so was Ahmed. The Iraqis were just dumfounded that I had deployed, leaving five children home with my husband, who is also a Marine, to care for them. After the story hit the streets, Ahmed said, "I have heard that there are many who are not happy with your story, but I am happy with it." [Laughs] He said, "A lot of people say, 'How dare this Mr. Amber say that women are the same? She wears pants!'" But, he was all cool when he said it. I really think he enjoyed the controversy. His wife was pregnant at the time with their first child and he'd sit down and ask me about things they were experiencing, like cravings, and he'd ask me, "Is this normal?" It was really cute because men and women don't talk there like Americans do. Not even close. It was like watching something so removed from our lives except, like I said, watching the women interact with their children which seems to be the same everywhere.

Vastly different from SSgt Amber's experience, SSgt Robin identifies the confidence in military might and her role in the execution of force she felt during her first tour in Iraq in 2003, what she calls "the original" and how training sometimes overrides issues of "leadership":

With the amount of aircraft we had there and the capability of the birds, I remember just feeling the confidence in ourselves and our gear. We had the potential to wipe Iraq off the face of the Earth. Whether we were allowed to use that capability or not remained to be seen and, of course, here we are years later still with the capability but adhering to moral standards, trying to save an innocent people while destroying the cancers within that region that are the true enemy…When the war kicked off and the Scud alarm would sound, we'd be sitting in the bunker in full MOP gear looking at each other, wondering how the hell we'd gotten there. I would realize that I was sitting in a bunker in the middle of the night in full MOP gear, and I didn't even realize that I'd woken up and gotten out of bed. That's how intrinsic Marine Corps training is. Your mind and body function on another level until your conscious mind can catch up. So, it was scary for about a week, then everything became passé. Our main concern after that was how freaking hot it was.

GySgt Tiffany describes the spheres she encounters and her extremely dangerous job in Iraq performed side-by-side with male Marines, but not as a sanctioned combatant:

We, as EOD, have a bounty on our head. I think it's a $40,000 or $50,000 bounty on our heads because we are the ones who are stopping what the insurgents are trying to do. They know that the IEDs *work* killing people. If you look at the statistics that is primarily what is killing our service people. EOD is stopping this. We're not door kickers. We're not going in and taking down the caches. We're not taking down the insurgents found in the buildings. That's for the direct infantry. That's not what I do, but I support the infantry when they find something. If they're on a convoy and they see something, they call me. There have been times when we'd get out there and it's IED after IED after IED spaced fifty or a hundred meters down the road, and we're clearing one and moving forward and clearing another and moving forward. That makes for some pretty long grueling days. I, personally, don't see a difference between me and those infantry guys. I'm out there potentially saving their lives. I've talked to a lot of male Marines about it and almost all say that, as long as females pull their weight and do what needs to be done and don't create a spectacle of themselves, the guys don't see a difference either.

CWO3 Leona describes how female Marines, taught to adapt and overcome, manage their spheres and are "kicking ass" in combat zones:

In the ordnance community we do what's called hot tube loading, which is basically loading rockets while the aircraft is spinning and it's a huge danger because of the static electricity created by the arc and everything. Plus, you've got all your weapons systems going, so you get guys loading rockets under there and it can be, well it is, *really* dangerous. So, I worked very hard on a lot of waivers over in Iraq, getting things approved for us to do because the publications just didn't cover them. We had the first Gulf War, and a lot of the publications stem from that era, but the situation this time is so much different, especially with the extended amount of time we've been there. So, many of the publications just don't cover what the actual operations are. Thus, when you get over there, you're in this gray area, but when you're dealing with ordnance and explosives and aircraft and people, you want a definitive answer, but there is no definitive answer. It's just not covered anywhere. So, I worked on a lot of waivers, like for arming and de-arming on certain spots. We had one arm/de-arm area across our airfield…and that airfield in Al Asad is *really* busy… We had like eight or nine different aircraft types there, if not more by now because the Air Force is there now, too. They had my ordnance Marines going across the airfield, an active runway, to arm and de-arm the aircraft so that they could come over to this side and re-fuel. I was, like, "What the fuck?" It just didn't make sense to me, and I put a lot of work into getting almost the impossible done. I got them to be able to arm and de-arm and refuel on the same spot, which is usually a no-no with fueling and stuff. There have been no accidents. Most of it was just red tape. You know, the military's been doing things a certain way for so long that they refuse to look forward even if there might be a better way. I knew the danger, but I firmly believed that the outcomes would justify the risk. I'm hearing good things back, still, so it's nice to know the risk and all the bullshit was worth it.

Unfortunately, with the victories in female leadership also come the tragedies. Captain Caroline describes duty in a FOB's Mortuary Affairs:

> The bodies would come up in so many different states. They'd be blown up, shot, drowned, beheaded. It was almost beyond belief what we'd see. We were just normal young adult Americans handling every day the most abused corpses anyone can even imagine. And, so many of them were our fellow Americans, our fellow Marines. Bodies of children, babies and their mothers. Old women and old men, grandmothers and grandfathers, whose only crime was shopping in the same market that they shopped in all their lives, except today was the day that an angry, selfish fundamentalist decided that innocent people needed to die to make him feel important. It was sickening, and there was no end to it. Some days were better than others, but no day was a "good" day in the morgue.

On another heart-wrenching note, SSgt Josephine describes one of her more challenging Marine Corps leadership experiences in Iraq:

> Seeing death like you do in Iraq just takes your breath away. There is absolutely *no* training that can really prepare you for what you see. What you can't do is let your Marines see that shock and disgust on your face. It's like with my kids. If they fall and split their head open or lop off a finger, I can't go to pieces. We stick a towel on the wound or find the lopped off body part, ice it, and head off calmly to the ER. They form their reaction from my reaction. If I don't go to pieces, they won't because they trust me to be in control. They trust me to lead them through the crisis. On my last deployment to Iraq, one of my Marines was severely burned in an IED incident. He had third degree burns over 80% of his body. When he arrived in Iraq, he was model pretty, about twenty years old, but after the accident he didn't even have a nose. He went from being a picture-perfect Marine to a horror movie villain in those few minutes on a street in Baghdad. We lost one Marine in that attack, and after they scraped him and a few other people up off the street, we went back to work. A few days later, I was able to go to the hospital and see him. The first thing I did was talk to the nurses and I asked them what to expect; the last I'd seen was just a charred form. They said it was really bad and told me that I'd have to wear a mask and all. I had to know what

I would see because I could *not* let him see shock on my face. I asked if there was a window. I wanted to see his reflection before I looked at him face-to-face. There was and I saw him. Then, I went to the ladies room and cried for, oh, about twenty minutes. I was so desperately sad because I still saw the beautiful young man, but anyone who had never known him would never be able to see that. On one of his hands, all of the fingers had been burned off. But, I got myself calm, and opened the door. The first thing you notice, of course, is the smell, but I was able to deal with that because we smell it damn near every day over there, and I stood in the door-way and said, "Well, hell. Some people will do *anything* to watch SportsCenter 24-hours a day." He couldn't talk above a whisper, but he said, "Hey, you came. I didn't think anybody would." It was so hard not to break down, but I couldn't allow it. It was too important for him to know that he was still that beautiful, brave young Marine that we loved so damn much. Sometimes, you do what you have to do to get through each moment, and there are a lot of moments like that in Iraq.

Captain Joy also describes a horrific tragedy in Iraq but lauds the training Marines receive to get through it all with minimal loss of life:

We deployed to Anbar province. Our primary role was convoy se-curity, running to Fallujah, Al Asad, Blue Diamond, and Rahmadi. Being in Iraq was the best experience I have had in the Marine Corps. It's exactly what Marines sign up to do. It's what we PT for. It's what we train for. I don't walk around with a combat ac-tion ribbon or anything, but seeing the things that your Marines do really puts a human face to the entire experience. I had 55 males and one female in my platoon. Seeing some of the really strong ones break, and seeing what you thought were the weaker or just quiet step up and take control, you really see what these people are made of and it's incredible. On numerous occasions, after the situation was resolved, I found myself thinking, "Wow. It works. All of the training really works." For example, Sergeant O'Hara was my first squad leader. He was the man, the myth. This platoon went where Sgt O'Hara told them to go. He was older. He'd been with the platoon the longest, and it didn't matter who came into or left my billet or the SNCO billet, Sgt O'Hara was the grounding force for the platoon and, as long as Sgt O'Hara was there, ev-eryone would follow him. Well, we were doing a dismount patrol looking for IEDs outside of Fallujah. A couple of weeks before,

Sgt O'Hara and I had been in Fallujah with the other platoon to be briefed on their mission of picking up the dead bodies in Fallujah after an "incident" because that was the mission that we'd be assuming so they could come home. So, when you're rolling through the city, if you're truck gets stuck, if you can't lift them, you get a pair of clippers and cut all the wires to get your truck through 'cause there's no power to the grid. So, on this particular dismount patrol, Sgt O'Hara was my assistant patrol leader and we came upon some downed lines, and he hopped out to lift the wires for the truck to go under it. He was electrocuted and died on the spot. That's when you really start to see the training in action. As soon as I said, "Hey, set up security. Do this. Do this." My PFCs and LCpls got on the radio, started coordinating things. They were the ones making it happen. The ones that were too close to Sgt O'Hara, the corporals and other sergeants, no kidding, didn't know what to do. They actually took their Kevlars off, their vests off, kneeled down next to his body and didn't know how to appropriately respond to the situation. Here we are, outside Fallujah, still on the street. I'm yelling at them to put their shit back on, to set up security! I didn't care how much they hated me for it. We had the rest of the group to worry about. We had no idea what kind of situation we were really in. Post security! Get that truck over there! Do this! Do that! Call for Medevac. Plus, we'd found an IED, so we were calling EOD at the same time. The PFCs and the LCPLs tracked everything on the radio and had it all set up. It was amazing because, when we were back on the road, I ticked it all off in my mind. We'd lost a charismatic leader. We'd set up our defense. We'd just called in EOD. We'd called in Medevac. We'd just gotten the hell out of there. Nobody else got hurt. It was successful but absolutely horrible.

It was a freak thing. It was an accident and it could have happened anywhere, but in Iraq, you don't know what's set up to screw your entire operation and your Marines. One thing can so easily lead to another that you can't assume that something like this was just "an accident." It could have been part of a much broader insurgent activity. It was one of the most horrible experiences I have ever had, but it also made me think about a lot of things. It made me realize how fleeting life really is. It also made me trust the Marine Corps so much more and the training we give our Marines. I really started to realize that this institution's been doing things a certain way for a very long time, and it works. Sure, you can't remove elements like chance, but the system really does work. It brought our platoon closer. Honestly, I was a 22-year-old female. It took a lot for those guys to say, "Hey, ma'am, what do you want us to do now?" Before, it would have been Sgt O'Hara that was giving the direction. It was a pattern. It was expected. He was an extremely strong leader and he was my right-hand man. As the senior sergeant, and in the absence of a SNCO, he *was* acting well within his billet to lead the junior Marines under my guidance. There was never a "power struggle" between us. This was the way things were supposed to work. Being a young female officer, I was truly blessed to have such a competent NCO under my command, and I won't ever forget that.

Captain Kai, a Military Police officer, describes a conflict between civilian and military thinking resulting in a demoralizing and operationally hazardous situation for her female military police dog handlers in Iraq:

Male Marines trust the females completely, but it's the media that has ruined the opportunity of females to excel in Iraq. I'll give you a perfect example. Infantry units going out on convoys or coordinated ops or whatever mission they were doing would request military working dog support from us. Two of our handlers, as allowed by policy, are female. So, we had experienced female handlers. I don't know if you're familiar with an operation called Operation Bad Karma. They had some press down there and had a couple of females involved in that coordinated search operation. The females gave an interview. This would have been sometime between March and October of 2005, and the females gave an interview explaining their role in the operation. Well, the press kind of twisted it a bit. The females weren't in the front of the stack, storming into the buildings, but they were there available to search female detainees. Apparently, there was a huge outcry in the States…I don't know how "huge" it really was…about women being on the front lines all of a sudden. So, in response, in Fallujah during that time period, the whole command got very sensitive to the issue of using women in "inappropriate" roles. I do remember a couple of Republican Congressmen asking questions about this particular operation. Up to that point, though, we'd been using the military working dogs with female handlers all over the place. We had females on convoys all over the place. There had never been a clear "front line" and "rear echelon" and there is no "support role" out there. There just wasn't. We were operating the way it was functional and it was *fine*, but I guess after this operation and the media attention it received and then the command's response, it called into question everything that had been working just fine before. So, we had units refusing to take our females and their dogs out. The gunnery sergeant who ran the working dog program was upset by that because that meant that his female Marines didn't get the same experience that the males did, and that the females had to do the same tedious jobs that the males now didn't have to do because they were on all the missions that the females were excluded from. It really hamstrung his ability to effectively run his small section. There were several times that I witnessed that the grunt units turned away working dog support because the handlers were chicks. These were units that the female Marines had worked with beautifully *before*, but now females were taboo because the commands were afraid of the public scrutiny of possibly having females in the wrong place. Nobody wanted a female to die or be injured or have some kind of incident or issue, and then have to answer the question, "Well, why was she out there?" While I was there, we really never did get the female working dog handlers back to the place they had been before the media put them in front of too many eyes. They were doing their jobs and doing them well, but their role was blown out of proportion by the media and everyone paid for it. The problem was *not* from the Marines, it was from the commands being scared of political pressure from back home. It's ironic that the operation was called "Bad Karma" because it really screwed up the balances that had been working with male and female Marines in Iraq to that point.

As evidenced by female Marines, managing contrary civilian and military expectations of their sex/gender inclusion in the Marine Corps is challenging. Combat scenarios, of course, embody the most extreme, but the charge of the military is the management of violence. In contemporary warfare, "combat exclusion" is a fallacy and destined to end *du jure* in 2016.[25] In reality, regardless of where she is assigned, a female Marine remains constantly on a different kind of battlefield. It is one of constantly shifting spheres in which she remains a warrior, a leader, a friend, a lover, a nurturer and the embodiment of so many other roles and expectations. This requires unique management of all aspects of her very *self* day-to-day, an experience that no civilian man or woman can even fathom, much less speak to. There is no going home at the end of a shift in combat; you're always still *there*. Through it all, female Marines remain closer to their male companions than to the rest of American society, a society that they believe does not understand them, yet judges them and imposes its contrary expectations upon them.[26] This often results in alienation and resentment with satisfaction derived largely from within the organization, from one's military family.

Endnotes

1   Lory Manning, "Military Women: Who They Are, What They Do, and Why it Matters," *The Women's Review of Books* 21, no. 5 *Women, War, and Peace* (February 2004): 7.

2   Brownson, "Battle for Equivalency," 21.

3   Alannah James, "The Myth of Jessica Lynch: Gender, Ethnicity, and Neo-Imperialism in the War on Terror," *On Politics* 6, no. 1 (Spring 2012).

4   Rosanna Hertz, "Guarding Against Women?: Responses of Military Men and their Wives to Gender Integration," *Journal of Contemporary Ethnography* 25 no. 2 (July 1996): 251-284.

5   Kanter, "Effects of Proportions," 972.

6   Ibid., 268-269.

7   King, "Women Warriors," 6.

8   Brownson, "Battle for Equivalency," 16.

9   Browne, *Co-ed Combat*, 165.

10 Schwartz and Rago, "Beyond Tokenism," 72.

11 Buss, *Evolution of Desire*, 264.

12 Ibid., 19-60.

13 Ibid., 265-266.

14 Brownson, "Rejecting Patriarchy," 4.

15 Moskos, "Toward a Postmodern Military," 22.

16 King, "Women Warriors," 1-2.

17 GySgt Scott Dunn, "Experimental Combat Unit to Test Integrating Female Marines," The Official Website of the United States Marine Corps, http://www.hqmc.marines.mil/News/NewsArticleDisplay/tabid/3488/Article/160476/experimental-combat-unit-to-test-integrating-female-marines.aspx.

18 Nix, "American Civil-Military Relations."

19 G.L.A. Harris, *Living Legends and Full Agency: Implications of Repealing the Combat Exclusion Policy* (Boca Raton, FL: Taylor & Francis Group, 2015): 296-297.

20 Williams, "Toward a Postmodern Military," 265.

21 Brownson, "Rejecting Patriarchy," 2.

22 Brownson, "Battle for Equivalency," 3.

23 Kanter, "Effects of Proportions," 972.

24 See Thomas Dietz, Tom R. Burns, and Federick H. Buttel. "Evolutionary Theory in Sociology: An Examination of Current Thinking." *Sociological Forum* 5 (1990): 162. The authors assert that group selection "holds that natural selection can act on entire *groups* as well as on individuals…genetic evolution could produce functional integration of the group and altruistic behavior that benefited the group despite the cost to the individual."

25 Julian E. Barnes and Dion Nissenbaum, "Combat Ban for Women to End", *The Wall Street Journal,* Politics and Policy, http://online.wsj.com/news/articles /SB10001424127887323539804578260123802564276.

26 Moskos, "Toward a Postmodern Military," 20.

Chapter Six

Professional Life in the FMF: The Enemy in Our Ranks

Professionalism and occupational competency are critical to military order and discipline, and women in the Marine Corps, simply because they are biologically different from males, often battle being marginalized, labeled, and stigmatized.[1] Appropriately and inherently gendered and passive, the the word "stigma" is the part of a flower's female reproductive organ (carpel) that receives the male pollen grains. Because they are sexually different from male Marines, females receive an overwhelming amount and variety of attention while facing unique challenges that would never be considered an "issue" for males.[2] The disparity in physical strength, a female's use or abuse of her sexuality to manipulate her environment, and the imperative to manage her body to make it seem less disruptive than it actually is are challenges unique to the military organization on levels unparalleled in the civilian community.[3] Sex and gender enactment potentially compromises female Marines' relationships with male Marines as well as other female Marines.[4]

Women remain barred from "combat" and the related accolades, which is likely to change officially in 2016 when females will integrate into combat military specialties.[5] As previously discussed, however, female Marines are exposed daily to the same dangers as combat troops largely due to insurgents' guerrilla tactics in Iraq and Afghanistan. Like their male counterparts, many meet the challenge and many die. In addition to emotional trauma, when in the field, a rigorous training environment, or a combat zone, especially, women must be prepared to manage menstruation, their hair, and other female-specific personal hygiene issues. In garrison, too, although men and women PT together as a unit, female Marines still have separate Physical Fitness Test standards with the Flexed Arm Hang substituted for pull-ups[6] and more time allowed for completion of the three-mile run. In contrast, standards for qualifying with the rifle, pistol, and swimming are sex-neutral. Females, however, tend to shoot poorly compared to their male counterparts. All of these reminders of their "Otherness" create unique challenges for female Marines.[7] Thus, the question arises: How can one be considered "equal" to a man, when one is constantly reminded of being "different than," "less than" or "more needy" than he is?[8] At what point do the allure of one's sex and the charm of one's femininity constitute a leadership failure and initiate a command crisis?

Many changes have occurred institutionally over time to reduce or eliminate sex discrimination and the stigma female Marines historically have struggled against since the AVF began integration.[9] However, one problem they still face is the fact that many within their ranks continue to reinforce the negative stereotypes by abusing or misusing their sexuality to manipulate individuals and the system overall. Accepting favors, feigning illness, becoming pregnant to shirk duty, seeking clerical assignments in lieu of working at "real" jobs, and copping out of field duty, professional schools, and deployment are only a few of the problems female Marines charge against their fellow females.[10] This chapter discusses an array of difficulties female Marines face daily simply because they are biologically female in an environment that often equates the feminine sex with weakness, manipulation, or incompetence.[11] It also discusses the anger and frustration female Marines feel toward other females who refuse to put their status of "Marine" before their status as "female."

Sexuality and Gender in the Land of Opportunity

Living in close proximity as Marines do, working, running, marching, playing, and sweating together, creates an environment in which an awareness of physicality dominates. In Boot Camp and OCS, bodily functions such as showering or using a field head are enacted as a group without the privacy of partitions. One female Marine remarked, "Where else but in the Marine Corps can you shower with 60 of your closest friends?" Living in squadbays, too, one becomes aware of the physical functioning of his or her platoon mates, even learning and living with their bodily habits and behaviors. Once in the Fleet, although not usually sleeping side-by-side with the opposite sex, female Marines are all too aware of the male presence and, obviously, the reverse is true as well. Perhaps familiarity does breed contempt and prompts social action. For example, historian Dee Brown asserts:

> A sociologist probably could make a good case for the proposition that the germinating point of women's rebellion against masculine authority was in the American frontier hotel. Here, for the first time in history, thousands of women came into close quarters with thousands of strange males; they saw the dominant sex as he really was, learned his empty secrets through thin canvas walls, and viewed his nakedness of his body and soul. In that locale may have been born the American woman's first doubts of the male's superiority over the female, doubts that would soon lead to demands for the right to vote, and for complete social and legal equality.[12]

The Necessity of Proximity & Physicality in the Marine Corps

Feminists' expectation that women in the military would revolutionize and "tame" the wild, misogynist organization missed the mark.[13] As presented in Chapter Two, it is not the demure and socially proper young women that seek a life in the Marine Corps. Rather, the females choosing to join the ranks are those eager to step outside social expectations and, literally, take a walk on the wild side.[14] The organization requires that female Marines adopt more masculine roles and become more active in very real-world warfighting processes. Any "taming" may merely result from female personalities and the different leadership styles they bring individually to the Corps.

Rather than taming, the proximity of the sexes and general living conditions often bring out the wild in both. SSgt Robin describes the situation in Iraq with boys and girls living together separately:

> It got really interesting when it got hot because everybody put their flaps up on their hooch. There was, literally, two feet between hooches and the guys that lived in the hooches next to ours would run around almost and sometimes actually naked. We'd tell 'em, "Hey, you don't see us running around naked. Put some

damn clothes on." We, of course, had to stop a couple of the girls in our hooch, too. They'd be prancing around in a sports bra and silky little PT shorts, and the female NCOs would say, "Uh…no. Girl, put some clothes on." They didn't care, though. Some of these girls were there to show off what they had and see what they could catch with it. One would put on her make-up every day. I was, like, "You're in the desert. It's 130-degrees. We're at war." It was crazy!

In contrast, SSgt Manda relates disappointment in being separated reluctantly from the male Marines in her unit, her comrades, while on a field exercise:

When we were in Pohang, South Korea, for Team Spirit, it was just me and one other female Marine, LCpl Post, who went advance party. For about two weeks, we all worked together and played together. Post and I lived in a corner of the guys' tent, mostly because we only had one heater. The guys hung a partition so we'd have a bit of privacy. I was sad, though, because when the main body arrived, a corpsman found LCpl Post and told her she was pregnant [determined from the physical exam urinalysis taken before deployment]. She was sent home on the next flight back to Okinawa. That left me alone with the boys, but staying with them with all those people around was a no-go. We'd all been fine together for two weeks, and I wanted to continue to live with the guys. We had it set up a pretty nice in the hootch with mink blankets and other cool Korean stuff. Instead, though, I was forced to live all the way across the pier with a bunch of other women from a mish-mash of units. We were all female, but other than that, I really don't think we had anything in common. I pitched a bitch to my SNCOIC [Staff Non-Commissioned Officer in Charge], but it didn't matter. If I tried to stay with the guys, we'd all get NJP [non-judicial punishment] or worse. The "girls" had to be banished to a tent behind concertina wire! It was embarrassing. So, when I wasn't on duty, I'd hang out with the guys as long as I could. Then, I'd trudge back through the bitter cold to the females' tent with the wire and a guard like a prison. I hated it because I'd left my friends, my family, all warm and happy together to go be with people I didn't know, much less like.

The experiences and perceptions of female Marines are highly sexualized, and rightly so. Their small numbers, less than 7% of the entire force, obviously draw attention to their presence and physical difference.[15] In addition, the Marine Corps environment demands adherence to a strict physical fitness regimen, weight standards, and body mass allowances. A Marine's very essence embodies good health, fitness, and even physical attractiveness. It is no accident that Marine Corps uniforms are tailored to display the physical charisma of the body within. As early as 1943, Elizabeth Culver recognized that the conspicuousness of uniformed groups brings privileges that entail very real responsibilities.[16]

Field exercises, barracks living, physical training, and so many other activities constantly juxtapose the male and female body in the same highly-charged space, often with mixed results. Sjoberg calls for a change in how the relationships between gender, war, and conflict are thought about, which might lead to a change in how those relationships work.[17] Her naïve expectation becomes problematic when actually considered in practice, though, with individual women acting in self-interest and not as a "woman cohort." More useful, Regina Titunik asserts that the myth of the macho military is overly simplistic and fundamentally flawed.[18] This research supports that fact and provides a foundation for individual women to negotiate positive relationships in the military through kinship and equivalence.

The Elusive Magic Elixir of Gender Neutrality

Interview data support the fact that female Marines from the AVF generation *before* much of the gender legislation was enacted experienced a *less* hostile and confused environment than that of today. Even if sexual harassment existed in deed, if not defined as such, it was dealt with as a leadership issue rather than as a legal one. Meštrović's identification of the fact that gender neutralization often involves making male/female encounters "foolproof" so that no one gets the "wrong idea" from the encounter serves as an excellent description of the current state of gender relations in the Marine Corps. He describes the postemotional requirements that social encounters between males and females must be made smooth, foolproof, problem-proof, and highly efficient. There can be no social space for the idiosyncratic emotional interplay that used to occur in mixed gender groups.[19] In the Marine Corps, however, this is simply not possible because, as determined in the previous chapters, every situation where a female is present is potentially sexually charged and negotiable.[20]

From 1stSgt Katrina's experience and numerous others interviewed, Meštrović's assessment can be applied to contemporary sex/gender relations in the Marine Corps:

> Women in the Marine Corps are *not* the problem. We are not "weak." I have a lot of males cry on my couch, too. If society expects us to be weak, we're a "problem" in the military. If society expects us to be strong in this environment, we're "dykes." This is just so wrong. The stupid rules to make us more unfeminine are stupid, too. When female Marines were expected to be feminine, like back before Tailhook, we didn't have nearly as many of the petty problems the leadership deals with today. So, now, they're going toward a unisex kind of attitude and we've got problems out the butt that we can't seem to handle through the methods that have served well in the past. Like de-sexing the females will make the males act less stupid? Maybe if we make everybody "green,"[21] they'll forget that they want to have sex? It's crazy thinking and I just don't understand it. If anything, all these rules only serve to reinforce the awareness that we're *all* sexual and instead of dealing with it rationally, we have to be separated and monitored... You can't take away sex from the human experience. You can only fix maturity with time and experience. These are fundamental human qualities that cannot be legislated.

This is a sentiment expressed repeatedly in the interviews, reinforcing the evolutionary psychology as well as the sociological literature.[22] Women Marines from the pre-Tailhook generation, although they hated the designation "Woman" before the title "Marine," expressed fewer concerns about sexual misconduct between the sexes. As 1stLt Rhonda says, "We all knew we were women. We looked like women. We acted like women. And, we were all cool with that. When they wanted us to be like men but at the same time shut off our sexuality is when the problems really began."

Virtually unheard of before the 1990s, "reverse" sexual harassment now manifests in the Marine Corps. As much as the males are "offended," the females involved are often dumbfounded. For example, GySgt Claire offers a story that illustrates the confusion about sexuality in today's Marine Corps occupational environment:

It's hot in the chow hall and a lot of times we work without [util-ity] blouses, just in our t-shirts. I actually had a male Marine tell my boss that he didn't appreciate seeing me with my breasts exposed when he comes through the line. He said that made him uncomfortable. It upset me at first, but then I had to laugh. It was just so stupid, but of course, in this modern Marine Corps, no one can feel "uncomfortable," so I dealt with that, too. It's like reverse sexual harassment, I guess…I don't even know what it is really.

Captain Sophia also relates a situation that confused her as well as the young man she "made uncomfortable":

As a female Marine officer, I do have luxuries that enlisted Ma-rines don't. I can also see how things can get out of hand. I, personally, have *not* seen sexual harassment, discrimination, or abuse. However, I actually had one young Marine tell me that I made him feel uncomfortable because, I guess, I winked when I was conversing with him. It shocked me because I wasn't even aware that I had moved my eyes at all. [Sighs] That was so weird and I really didn't know what to do with it.

Thus, in the present postemotional Marine Corps, both sexes are obviously confused about appropriate versus inappropriate be-havior relating to the opposite sex. In the midst of all of the confu-sion, however, female Marines continue to be categorized as one "type of girl" or another.

Role Assumption, Stigma, & Effect: Bitch, Slut, Lesbian, or Equivalent

Many anthropologists, evolutionary psychologists, and so-ciologists agree that humans categorize experiences in attempt to cognitively understand as well as act and react comfortably in our reality. For example, anthropologist James Brain relates the human requirement to categorize, an idea developed by ethnologist Arnold Van Gennep as:

[A] general human concern with categorization, with order and disorder – with anxiety because of the inability to impose order and, arising out of this, the attribution of danger and/or power to persons and things that are not readily put into the categories of a particular culture. They fall into the anomalous position of being marginal or liminal, and potentially polluting.[23]

Thus, per Brain and in agreement with Kanter, it only makes sense that humans will categorize individuals in reference to the norms, expectations, and goals of the group. The Marine Corps is no exception. In fact, one might accurately assert that the Marine Corps experience magnifies this tendency.[24] For example, Sgt Justine describes role prioritization and the ensuing categorization as an on-going process for female Marines:

> I think a female has to decide what role she's going to put as a priority. Are you going to be a Marine? Or, are you going to be what society expects a female to be, weaker and demure, and all those other things that make men more important than women? A female Marine has to decide this over and over. Different situations require that a female Marine makes different choices between those two questions. For men it's simple. If he's a man, being a Marine isn't a stretch. He's just being a very manly and disciplined man. For a female, though, she has to manage her abilities and her roles, trying to guess at the expected outcomes of those choices. Like, if a female decides to try to smoke the time on the three mile run on a PFT, she has made a choice to be ultra-competitive which is viewed as "masculine." If she's successful, she receives accolades from the "good" male Marines while the "not so good" male Marines resent her for making them look bad. Unless she is well-respected by the female Marines in her group and is already viewed as a role model, she is now viewed as a traitor to the female Marines who will never be viewed positively by the males because she has now made their attempts at being "good enough" a mockery.[25] None of these decisions are easy. I think some are not even acknowledged in a woman's thought process when she makes a decision. However, the outcomes are *real* and become part of her reputation, like it or not.

Although acknowledging a female Marine's agency in constructing her reputation, SSgt Lacey includes the assertion that physical limitations must also be included in the equation:

> A female Marine has to decide what *kind* of attention it is that she wants. You can choose the negative attention and be a problem to yourself and everyone around you, which undermines everything the Marine Corps is all about, or you can choose to be a

motivated and inspirational leader. This is a really tough decision and, honestly, I think for many, many women it's just out of their hands because they couldn't pull it off even if they *wanted* to be a high-quality female Marine. If you can't run…forget it. If you can't hump….forget it. If you can't lead and make appropriate decisions…forget it. Maybe for a female like this, being a "slut" is the only way she can get by and *not* be invisible. So, maybe they're not just being "bad" Marines; maybe these females are looking for validation in their own right. Choosing to join the Marines may look like a "good" opportunity, but it may just be a poor choice. Once you're here, you really have to make it really work, and some just have no chance of making it work with the skill set they bring to it.

Similarly, Captain Kami describes the issue of role management as a leadership challenge as well as a personal issue:

My Master Sergeant says that I need to be more strict with the Marines, but I tell him, "That's why you're here, Top. *You* be the hard ass. *I'll* play the good guy." Isn't that the difference between the officer corps and the SNCO leadership? It's particularly difficult for a female, too, because if I'm at all pleasant and approachable, I open myself up to questions about my moral character. Am I being lax or "loose"? Am I trying to seduce my male Marines? Am I treating all of my Marines fairly? If I'm hard, I am considered a bitch and "difficult to work with." Even better, they'll say that I "have an ax to grind" or that I am a "man hater." These questions simply aren't relevant for male officers. Oh, it can get really crazy and spiral out of control in an amazingly short period of time. In reality, I'm a big goofball. I like to have a good time, but it is so difficult to manage all of the relationships and expectations and not be too much of one "thing" and not enough of the other "thing." That's been the biggest challenge of my career, not the PT but finding that delicate balance between "bitch" and "mentor" and "professional" that is what the Marine Corps expects in its officers. It's not easy, but I do enjoy the challenge. What I find, too, is extremes. People, Marines, either really like me or they hate me. There is no in between… Being a female in the Marine Corps is like being a rock star. You can be famous or you can be notorious. Sometimes, you don't get to choose your status; it's chosen for you. I have to be so much more diligent in what I do because I know I'm constantly being watched…I actually had one enlisted Marine tell me when he was trashed at Bosses' Night, "Ma'am,

I've got to tell you. My first impression of you was that you are a dominatrix." I was floored, but then I thought about it and thought maybe that's not so far off. I have dark hair with a widow's peak and very pale skin which is a stark contrast. I am taller than the average female. I am a female Marine. Hmmm…maybe he was on to something. [Laughs] Since then, though, I have worried and thought about "softening" my appearance.

Within the equation of making sex/gender relations "work" in the Marine Corps, the variables have only become more confounded over the years. During the Free a Man to Fight generation, women were not commanded by men, nor were they expected to command men. During this time, three basic categories of female Marines originated: the lesbian, because of the desire of a woman to pursue a masculine career, the "loose woman" (a.k.a. the slut), and the "nice" girl who conformed to societal expectations. [26] This categorization occurred in a time when male and female Marines were segregated, separate and inherently unequal, in almost every aspect of their experience from initial training and throughout their occupational service.

Kanter identifies similar categories in her influential study of tokens in the 1970s: Mother, Seductress, and Pet.[27] Since the Free a Man to Fight generation and Kanter's work, however, I assert that the three previous categories expanded into four with slight variation since, and likely due to, the All-Volunteer Force generation. Christine Williams rightly stated in 1989: "…female marines are to some extent flattered by men's attention and constant notice. Those who object to this treatment are considered hostile and bitter by their peers because, by refusing to accept this interpretation of men's behavior toward women, these women challenge the very grounds for other women's acceptance of the status quo."[28] This awareness of her biological sex and its impact prompts a female Marine to manage her public identity and the perception of herself by those around her. The female Marine described by Williams would appropriately be placed in the relatively new "Bitch" category, the I-know-I-got-it-but-you-can't-have-it category. Her sex embodies power and she remains aloof. Also due to the changes of females' roles in the Marine Corps from passive bystander to much more active participation, the "nice girl" category evolved into "Equivalent." The modern Marine

Corps, it seems, is no longer a place for "nice girls," but respected peers who exhibit kinship commitment and ideals about warmaking processes are welcome as "equivalent" to males. Contemporary stereotyping categories validated by the data herein, Bitch, Slut/Seductress, Lesbian, and Equivalent, demand further investigation and clarification.

Categories & the Stigmatization of Contemporary Female Marines

The Bitch model reflects professionalism, strength, and competency, and refusal to adopt a nurturing or accommodating feminine role unless it is part of her professional duties. The latter she likely performs perfunctorily and without empathy. Her behavior can be viewed positively (in support of the group's mission) or negatively (when considered selfish, undermining the group's mission). She remains separate and aloof. The antithesis of this role is Kanter's Mother which, I argue, is archaic with no place in the modern Marine Corps organization.

The Slut/Seductress model remains a fixture throughout time in the military.[29] This female may or may not embody strength and competency. The evidence suggests, however, that sluts rely on employment of their physical sex and feminine wiles because they do not manifest the professionalism, strength, or competency of the Bitch or Equivalent. The Slut/Seductress' behavior is disruptive to relationships and the organization because she flaunts and uses her sexuality to gain attention and favors.[30] Men may appreciate her free dispersal of sexual favors, but her desirability as a long-term mate is questionable. Men's preference for physically attractive mates is a species-wide psychological mechanism that transcends culture; however, from an evolutionary perspective, chastity is, in fact, a necessary virtue.[31] Further, when her behavior is environmentally disruptive, at best, or she is a mate stealer, at worst, she tears at the communal social fabric.[32]

Traditionally, the Lesbian model rejects biological and socially defined roles, behaviors, and expectations. As previously discussed, the human brain strives to categorize. Although advances have been made in American society to educate about alternative gender expressions and encourage embracement of diversity, Lesbians defy categorization within the military organization. Although comprised of a physical, sexed body, they are not reflective as male or female based upon this. They are not neuter because that embodies an absence of sexuality. Thus, the Lesbian generates confusion.

The Equivalent model embodies kinship with her peers through commitment to common values, goals, and behaviors reflecting male-female co-evolution as an historical emergent from conditional interactions throughout human history.[33] Equivalency encompasses relationships between and among individuals without imposing the expectation that anyone, male or female, must transcend their biological or social nature to be relevant to the community's, individuals,' and mission's success. It is pragmatic, a shared and enhanced commitment to Marine Corps ideals.[34]

The Bitch

The "Bitch" label in the Marine Corps works positively or negatively. The negative association reflects on a female who refuses to be nurturing or otherwise feminine, such as actually adopting a masculine aggressive persona. As negative characters, the females in this category are viewed as selfish and often as "posers," insincere in their role as a female Marine, socially and/or sexually. These females are generally not trusted, unpopular, and sometimes even dangerous. For example, Sgt Justine, a corrections MP, describes one of the young females with whom she works in the brig at Camp Lejeune:

> There's a girl who hasn't been here very long; she's just out of corrections school really. She's been overheard saying, "A prisoner's a prisoner. They're all nasty mother fuckers to me." She's one of those women that has to put on this tough bitch act to feel secure about her status with the prisoners. It's her defense, but the attitude undermines what we're really here to do, which is rehabilitation. The prisoners, too, see right through the act. They may have made bad choices, but most of them aren't stupid. They can almost smell false bravado and, as Marines, I'm sure there's a part of them that wants to pull the punk's card. So, when she thinks she's taking a hard line to earn respect, she's really just setting herself up for ridicule from her peers and possibly danger from the prisoners.

Positively, however, "Bitch" also applies to extremely competent females who are simply getting the job done. This "Bitch" status is usually temporarily inflammatory as the Marines around her begin to respect and trust her as long as her perceived bitchiness as fostering the greater good. In the cases where her evidenced leadership skills override the presumption that she is being "bitchy" solely out of meanness or insecurity, a female Marine asserts her authority and earns respect for her command of the situation and of her Marines. For example, SSgt Tomasa describes her leadership and authority in relation to her junior Marines:

> If my watch pissed me off, toward the end of the shift, they'd be scrubbing to get everything clean to go home. They'd say, "Everything's clean." I'd say, "No. I don't think it is." I'd get trash cans full of steaming soapy water and just kick them over. "Remember earlier when I *told* y'all not to be fooling around? I meant it. Get busy. I ain't got nowhere to go, so scrub, scrub, scrub." I'd hear them saying, "Oh, she's such a *bitch*." I don't care. The next time I tell them to do something or to stop some foolishness, they do it. And, it's funny, too, because they'll train up the newbies, too, before they can screw it up for the ones who've been there for awhile.

Likewise, GySgt Tiffany also has no qualms about describing herself as a Bitch:

> When I was in Boot Camp, we were told that we could expect to be labeled as one of two categories: We would either be labeled as a whore who was sleeping with everybody, or be labeled a dyke because we weren't. I raised my hand and said. "I've got another one. I'm a bitch because I'm neither of those." I have gained the respect of the guys that I work with because I am not going to whine. I am not going to pull punches. Every single female in the Marine Corps knows that they have that card [sexual harassment], that ace in the hole, and anyone who says they don't is a liar. That's why a lot of guys don't respect females or are leery about being around females; they know that, if her *mood* changes, she can pull that card and they might be screwed beyond all recognition. Before, there was a culture of "blame the victim," but now it's open season on the males. I have a real problem with any female that uses that card as an offensive tactic because it ruins the relationships all of us other females have worked so hard to establish. In a few moments, years of camaraderie-building is destroyed…I'm a true, honest person, and I think that's why a lot of people may not like me. I'm not going to play little social games designed to

make you feel better about yourself. Maybe you need a therapist. I dunno. I have better things to do. I am honest to a fault, but you will never question how I feel about you. If you want to cry because I don't like you, you are probably placing way too much importance on my opinion. My advice if you're unhappy: Seek a second opinion, and just don't worry about me.

GySgt Claire prefers to describe herself as "assertive" rather than a Bitch:

I am a strong minded black woman and that's gotten me in trouble a lot. [Laughs] I came to the Fleet as a private. I was older, having gone to college for a year-and-a-half. In my family, the women rule. I spoke my mind. I had no tact, and that got me in trouble a lot, too. I had one staff sergeant tell me, "You need to stop telling me 'Fuck you,' with your eyes." I said, "Well, Staff Sergeant, do you want me to say it with my mouth?" He'd been yelling at me like I was his kid. He got crazy, and that made me mad. The other females were like, "Hi, Staff Sergeant," and bat their eyes at him. I wasn't like that. I was there to do a job and to do it well. If I messed up, I'd fix it. I wasn't there to get him to like me, so it irritated me when they acted like that. I can't stand when the females are acting silly to get attention. So, those girls behaving like that feed the egos of the males who then expect every other female to treat him like he's a god or something. I was known as shit-hot because I could run, but I was also known for my attitude. I was known as a bitch, but I prefer "assertive." Maybe it all balanced out? I dunno. [Laughs]

The female Marine Bitches interviewed all expressed their sincerity of intent and confidence in their ascribed and assumed roles. Many expressed dismay at feeling like they had to be more of a hard ass than they normally would have liked just to prove a point to the Marines around them, either to prevent themselves from being dismissed as a female by those above them or to command respect from those below. In sum, their tactics seem to work when sincere and even cross over into Equivalent status. The "bad Bitches," however, are easily identified as posers, weak, and hiding behind bravado and, as such, often do not earn the respect of anyone around them.

The Slut/Seductress

To explain why casual sex is taboo, Buss asserts that from an evolutionary standpoint, the chastity of females is more highly valued than male chastity by humans because it ensures correct paternity and is likely an indicator of marital fidelity, also directly linked to paternity. In this situation, a male has confidence that the children he has committed to are, in fact, his and he is not wasting his resources on another man's progeny.[35] From a sociological perspective but asserting exactly the same concept, Blau describes the loss of value of a commodity through its ease of attainment, and since the sexuality of women in most social structures is perceived as "owned" by the male who "owns" the woman, her sexual value declines as she dispenses her sexual favors indiscriminately. "The interest of girls in protecting the value of sexual favors against depreciation gives rise to the social pressure among girls not to grant the favors readily,"[36] and promiscuous behavior is viewed by both men *and* women as deviant.

In defiance of the assertions of Buss and Blau, we cannot deny that we live in a "hook up" culture where casual sex is largely accepted. However, the deviant status of the slutty female Marine in the military emerges from several very real and disruptive situations: pursuing gratuitous or manipulative sex, having sex with married men, and promiscuity.[37] According to evolutionary psychology and traditional social norms in the vast majority of cultures, a woman's sexuality is controlled by the "owning" male, which is further supported by the kinship group because of its value for the production of future generations of proven reproductively fit humans.[38] It is therefore a social commodity to be guarded and granted only in socially ordained circumstances, such as marriage.[39] In addition, when the mating "rules" are rejected, the expectations of others are thrust into a quandary. As Buss says, "Many shun the promiscuous and scorn the unfaithful because they often interfere with our own sexual strategies."[40] In the Marine Corps, sexual indiscretion can also devastate good order and discipline.

It is not surprising, however, that men are often willing recipients of the wanton's favors. For women, the behavior is labeled deviant because it devalues female sexuality in general and, perhaps, because easy access to such a desired commodity undermines the more guarded commodity of others. Potential mates are busy dallying with the easy girls, while the Nice Girls/Equivalents wait for their long-term mating attention or simply choose another acceptable mate while waiting. In the words of Sgt Reese:

> It was interesting because there were situations where the guys sought us [females] out, like for intramurals where they'd be competing and you had to have a co-ed team, like for weight lifting, volleyball, relays, wrestling, and the like. So, they'd come to us, the "good" females and ask us to be on their teams because we were strong or fast, depending what the sport was, and they knew we'd be dedicated if we agreed to be on a team. The slutty girls absolutely *hated* that and hated *us* because the attention wasn't on them anymore. They *hated* the fact that the guys wanted to spend time with us, training and competing and whatever, because we were competent instead of easy. Sometimes they'd try to start a nasty rumor about one of us, but it just wouldn't fly. The guys would be like, "Yeah, whatever. That whore's only good with a mattress strapped to her back and is jealous." It was incredible hearing them say things like that, the guys actually speaking to the distinction between us and them. Those same guys, I would bet, would still fuck the slutty girls, but I don't think that would affect the way they felt about us, the "good" female Marines as peers. I think the distinction was just that sharp, and therefore easy enough for them to take advantage of the one, the sluts, while accepting the others, like me, the females that wanted and deserved their respect and friendship.

Sgt Winnie provides an illustration of a female Marine truly "whoring for the Corps":

> There was a female Gunny that arrived a few months before I left [recruiting duty]. This wasn't her first tour as a recruiter and she asked me, "Do you want to know what I do to be successful at recruiting?" I almost said, "No," because I had a feeling about what she was going to tell me, but I let her tell me anyway. She said, "Out here, I shorten up my skirt. I wear visible thigh-highs. I leave open the top button on my shirt, and I got a boob job the first time around." So, basically, she whored herself out to recruit for the Marine Corps. Maybe she did it to make herself, personally, feel better. Maybe she did it to enhance her career. I really don't know. All I do know is that I was absolutely disgusted by what she said and what she was doing. She misrepresented over two hundred years of valor, *plus* how many dedicated men and women to

"sell" the Marine Corps in this manner? I love the Marine Corps, and this, I believe, is totally the *wrong* way to entice young men to enlist, and what does that say about females in the Corps, that we whore ourselves out to secure male recruits? It's sickening. She was a SNCO, but she perpetuated the misconception that women are only important as sexual beings, that we can't be *real* people or *real* Marines because all we have is our sexuality and nothing else to offer humanity. It's just damn sad. I stayed away from her after that but, of course, she made me look bad because she started writing contracts from day-one when it looked like I wasn't able to do a damn thing as a recruiter.

First Lieutenant Brieta, an admitted slut, describes why she did the slutty things she did as a young enlisted Marine:

I loved being a nihilistic little nymphomaniac. [Laughs] Really. I am not going to deny it. I absolutely *loved* being able to do whatever I wanted with whomever I wanted to do it with. I never felt so free and happy as when I was seducing someone or gaining another sexual experience. I guess I fit completely the WM-slut stereotype, but I chose to be that way and didn't care. The men were fucking everything they wanted to and so was I. [Pauses] I had some kind of curiosity that drove me, and allowed me, to have sex with as many men as I wanted to without any sense of guilt. I wasn't careful. I, we, didn't use condoms. My goal was to gain experiences primarily, but not solely sexual, and I did that. It was crazy and exciting and fun. And, sometimes it was scary when things got out of hand, but somehow, I loved and still appreciate all that I experienced. I can't even guess how many sexual partners I have had, but I was so blessed I never contracted an STD and I do *not* have HIV. [Laughs] I think that when I am old, I will regret things that I had the opportunity to do but didn't, but I *will* die with the satisfaction of having done the things I did. I have no regrets, except not doing some of the things that I now realize I should have done to complete my quest for experiences. I was and still am a junkie. As a wife and mother, though, I can't do the things I used to do. I play by the rules of marriage, but I am, however, the kind of woman that should never marry, never be tied down to one man. I do know this. And, although, I will never cheat on my husband, physically, I continue to yearn for sex and physical intimacy with random other men. It's funny, but I don't want to own them or possess them like women feel about their husbands and lovers. I just want to have that fleeting, or maybe longer, physical relationship but without the ties and expectations. I certainly don't want to wipe their kids' noses or wash their dirty dishes and clothes. What woman would *choose* to do that except for security? I don't see the fairness of the trade-off. Honesty, I love men and I just want to fuck them passionately and learn what makes the sexual experience as intense as it can be for them as well as for me. No man will ever own me, and I love that.

It is obvious that some female Marines revel in their sexuality and use or abuse it for a variety of reasons. Regardless of what those reasons are on an individual basis, their behavior potentially reflects on the entire group.[41] Consequently, when a female Marine arrives at a new duty station, one of the first questions seemingly on every male Marine's mind is whether or not she's sexually available and, of course, promiscuous. In addition, females not engaging in slutty behaviors are also assigned the label, often forcing them to adopt even more conservative behaviors than they would like in order to reduce the damage that a bad reputation can have in the Marine Corps. Once labeled a slut, the label is difficult, if not impossible, to shed.

The Lesbian

Being stigmatized as a lesbian is disturbing to many hetero-sexual female Marines, primarily because this label, too, presumes to steal their true sexuality from them. Matt Ridley puts the situation in perspective from an evolutionary standpoint:

It is sometimes hard even for biologists to remember that sex is merely a joint venture. The process of choosing somebody to have sex with, which used to be known as falling in love, is mysteri-ous, cerebral, and highly selective…The urge to have sex is in us because we are all descended from people who had an urge to have sex with each other; those that felt no urge left behind no descendents.[42]

Evolution obviously favors male-female coitus to procreate. Many lesbian females serve with distinction, but the labeling of heterosexual female Marines as "lesbian" negatively impacts their lives. To be targeted as a lesbian indicates to other humans that one is not interested in males, a concept considered by many as offensive on a biological as well as social level. Sgt Jeanne describes a challenging situation she faced while on recruiting duty in Texas:

When I went to get my truck fixed, the mechanic who had just graduated from the local high school commented to me about the couple of bumper stickers I had on it. One was "I fling poo" and I don't even remember the other one. But, he asked me if I was a lesbian since that's what I was advertising on my vehicle. I couldn't believe that he had the balls to say that to me! I was, like, "Whatever" and just went on. I didn't dare get into it with him. I was a recruiter. My job was selling the Marine Corps, but these people in this town were….I don't even know how to describe it. I guess the movie "Deliverance" comes to mind. They were just so backward and absolutely convinced that they knew it all. So, this guy apparently passed his valuable information on to all the other young people in the area because soon it was well-known that I was a lesbian and that all the females should beware because I would try to convert them and the guys needed to beware, too, so that I couldn't get at their girlfriends. One of the girls in the high school, I actually took out to lunch because she expressed interest in enlisting. But, toward the end of the meeting, after I had told her all I really could, I asked her if she was interested and she said, "Oh, God, *no*. I just wanted to see how a lesbian would try to pick me up." I could not believe it! Just when I thought I had experienced all the bullshit they could say against me, this was just unreal. So, after that, I started carting around a male companion wherever I went. I know it sounds stupid, but I was in sales, and impression and perception is everything. I was a Marine with a security blanket, *not* for my physical protection, but to protect my reputation as a heterosexual female.

Another challenge posed by the stigma of lesbianism is the actual or perceived predatory nature of lesbian women in the Marine Corps. SSgt Lindsey describes an event at MCRD San Diego while she was stationed there in the early 1980s:

Myself and some other women Marines were witnesses to some homosexual acts. At one point there were so many lesbians there, they had almost an entire wing of the barracks. What they had started to do was when a young private would come in straight out of Boot Camp, they would be the first ones to befriend her, invite her to hang out with them, and the next thing you know, you've got this 18-year-old girl in with the lesbian crowd. They were actively recruiting young females into lesbianism. No one was really saying anything about it. People make their own choices, you know. But, they got to a point where they were really aggressive with it. There was one girl who'd never had a boyfriend. She was 18- or 19-years-old and she reported it because they were making advances at her. Once she reported it to her sergeant, CID came in to investigate it and it became a big thing. She wore a wire, and a lesbian girl tried to rape her and was arrested. There were about three or four of us who were witnesses at her court-marshal. During that time it was a big thing because it leaked out into the community. People in San Diego were outraged because there were lesbians in the military! Then, you had another group asking why the government was attacking their lifestyle. It was a really big mess. These women were just as bad as any male predator I have ever heard of. Had they not been women, physically, they would have been men, and, maybe worse than men because they felt like they had something to prove by not having male genitals to seduce these other women. They did the same things a man would do: calling you at work, showing up at your door, mad when you went out with someone else, bringing you alcohol saying, "We just want to have a drink with you." But, you knew what it was they wanted. They were just as predatory as the worst male. That was the most serious incident, but there were a few other situations with lesbian women, but it never became the incident that we experienced at MCRD San Diego. There are always incidents in the Marine Corps where you have a guy who can't keep his hands to himself or he can't keep his comments to himself, or just plain ol' didn't want women in the Marine Corps. But, these women... it was just so unexpected and just so much more *wrong*.

Female Marines who are not homosexual generally are ambivalent towards lesbian Marines and many are good friends. They are troubled, however, when they, like the sluts, become disruptive and their behavior reflects negatively on all females as a group. For example, Captain Paula recalls being the victim of lesbian sexual harassment by a female early in her career:

After five years in the Marine Corps and never once experiencing sexual harassment from a male Marine, I had to file a harassment complaint against a lesbian boss who developed feelings for me and pretty much made my life miserable. She called me incessantly. I stopped taking her calls because a third of the time, they were business but the rest were personal. She was very needy, very lonely. Honestly, I can't imagine being a lesbian in the Marine Corps because you'd have to suppress so much of who you are. Your personal life would have to be kept a secret and I don't know how they live with that. I think it makes them a little crazy, and I think it definitely drove this woman crazy. It was incessant calling and she never wanted me to leave work. Eventually, she got drunk one evening and expressed her romantic affections and made a pass at me. We talked about it in person, alone, and I just told her that I wasn't into that at all. After that evening, I spoke with her again and told her that I couldn't work for her anymore. Obviously, I didn't want her writing reports on me anymore. It was something out of a bad EO "This is Sexual Harassment" video when she told me that I had asked for it, that I had flirted with her…I had led her on. So, I immediately took it to the command and they handled it. To this day, though, I feel like I am a victim of "Don't ask. Don't tell."[43] When I came into the Marine Corps, I was very liberal in my thinking. My opinion was that gays should be allowed to serve openly. What I experienced, though, is a really scary thing. President Clinton's policy endorse[d] lying about who a Marine is, fundamentally, if you're telling a person to hide their sexuality, which is one of the most basic things that makes a person who he/she is. It completely belies the Marine Corps core values of "Honor, Courage, and Commitment." You can't be a Marine and be confident in your own skin if you're living a lie, even if it's a lie that the President of the United States has authorized.

Like so many issues in the Marine Corps, the issue of homosexuality/lesbianism waxes and wanes as a focus of media and, thus, institutional attention. Although occasionally disruptive as evidenced above, for the most part Marines are aware of homosexual Marines, out or not, and are unconcerned about their presence and lifestyle until someone crosses a professional boundary or invades one's personal life.

The Equivalent

Female Marines recognize the disruption that their sexuality, their mere presence, creates in the Marine Corps.[44] The females who strive for excellence also recognize the need to manage that sexuality to neutralize the working environment for the benefit of themselves as well as their male counterparts.[45] Many females profess that this is *not* a compromise of their femininity, but simply a management tool to enact the various aspects of their *selves* appropriate to the situation. Anthropologist James Brain quotes Freud who asserted in 1905 that "observation shows that in human beings pure masculinity or femininity is not to be found either in a psychological or a biological sense. Every individual on the contrary displays a mixture of the character traits belonging to his own and the opposite sex."[46] This concept permeates this research. Successful female Marines find a way to strike a balance to "fit in" in this masculine subculture while also maintaining the aspects of "self" that are feminine.

Maj Lauren offers her opinion having seen changes in the Marine Corps over three decades, from the early days of the All-Volunteer Force to today expressing her sense of kinship and equivalency:

> I always felt like I was with a group of brothers. We were like family and if they screwed with me, I screwed with them right back. I honestly think that we've over-trained in some areas. I tell my females today, "Hey, this is *not* an Equal Opportunity issue, this is a leadership issue." Somehow, I think the Marine Corps has lost sight of which is which, and that's when it gets so confused in everyone else's minds...there were always issues as a female coming into these all-male units or doing these new things that females hadn't done before. It was never easy, and there was never anyone to really work through it with me. I was totally on my own, making it up as I went along. I dealt with the guys in a friendly, but no-bullshit-allowed, kind of manner. I was respected because I wouldn't allow myself to be categorized as weak or incompetent. As time went on, we did have wimpy females. We had one girl later on who had to have her toolbox carried out to the bird she was assigned to work on. That pissed me off. I was working very hard to prove to the guys that we were good enough and strong enough to be their equals, and then we had this chick check in and embarrass us. That upset me greatly. I think she didn't have a clue about maintaining a positive reputation of females in the Marine Corps, and I think the same is true today with these kids coming in. They don't see the "big picture" and their place in it.

SgtMaj Brittany describes her kinship experience and her pleasure with what she believed to be the absolute best of both worlds:

> It was never said to my face that I was a bitch or a slut or a whore.

In my first band, I did ask one of my male friends why I wasn't experiencing those problems that I saw the other young women struggling with. He said, "You're just our sister. You don't flaunt your sexuality at us. You're just you and you don't expect anything from us other than our respect and friendship, and you earn both every day." I don't need another person or a group of people to define who I am. I am very comfortable within my own self. I can be by myself and be perfectly happy. I can be with a group of people and be perfectly happy, but I have no expectations from that group and that group won't define me. I thought it was neat that the guys saw me as a sister, a friend, not someone who's going to try to use them or get over on them or play them. That's not who I am and they know that, and that's why I am accepted and successful like I am. That's a very good feeling, to be free and content without those expectations and labels.

Prior enlisted LtCol Abigail, a Marine Corps attorney, describes how her infiltration of the male domain began as an enlisted reservist and then continued as an officer at TBS:

In my reserve unit, there were 120 men and three women. I was always in the field. I was always one of the guys. At TBS, there were 17 guys in my squad and me. Initially, they were much more apprehensive than I was. The Company Commander took us aside and said, "We're not going to treat you any differently than the men with the exception of two or three things that, by Marine Corps Order, we were not allowed to do, like the live fire ambush, pugil sticks, and I can't remember what else. We ended up doing the live fire ambush anyway. There's no way I was gonna miss that! [Laughs] In our first peer evaluation, I was toward the bottom of the middle third, maybe top of the bottom third. It was all based upon garrison activities, but I was disappointed that I was ranked so low overall. Our SPC was a great guy. He was an infantry guy who never worked around women. He said that he didn't think that the women, in general, were getting a fair shake in the evaluation process.[47] So, when we started going to the field and I started doing more and more things, my male peers were pretty shocked and amazed at the things I was actually able to do and that I could actually hang with them. We had our final "war exercise," which was a nine-day evolution and I was the company radio operator. I had to carry all this gear for the radio and a grenade launcher and grenades, plus my pack and rifle. So, I was weighted down as much if not more than any of the guys out there. It was easily over half my body weight. Later, when I was out in the Fleet and ran into him, I asked the SPC what was up with loading me down with so much gear and he said that they had to do it for the guys to realize what I was really capable of and that, until I did that, they would never really see how competent I was. After that exercise, I was ranked #3 in the company with comments from some guys that were going to be infantry officers that said, "I'd take her with me anywhere." *That* said so much! TBS is a cut-throat business, and to have male soon-to-be-grunt-commanders throwing their opinion and their reputation behind a female was really something, and it did *not* go unnoticed. I was a minor celebrity, which was *not* what I wanted, but you take the accolades any way you can get them in this Marine Corps. Physically, if you can perform, whether you're male or female, that's the first thing you're evaluated on. You cannot be weak and be successful. That's just not going to happen, and I proved that I was one of them.

1stSgt Jackie describes promoting a Marine Corps "ooh-rah" mentality among her Marines, disbursers, in Yuma, Arizona, always putting the Marine Corps team before the individual:

I did make disbursing fun for my Marines. Yeah, we worked in an office, but I made sure that we did "Marine things." People don't join the Marine Corps to sit behind a desk! Any other branch of the service, maybe, but Marines focus on being Marines *first* and then your job is next. That's a challenge even today, keeping Marines focused on the fact that they're Marines, first. I was bored, so I did a lot with the disbursers. We'd go out and do the combat conditioning course, and we'd go on humps. We were an anomaly because, one, we were in the Airwing and two, we're disbursers. But they loved it. It set us apart because we were being Marines *and* doing our jobs, and loving all of it. When I was in the hospital giving birth to my son, a few of them came to visit me and they'd say, "Hey, cute kid. Oh, by the way, there was a WTI on board and they asked us if we were grunts! And, we said, 'No, we're disbursers from the air station.' [Laughs] It was *SO* cool!" Those guys couldn't wait for me to get back from maternity leave to tell me that they'd been confused for grunts. Seeing how happy they were made my assignment there with them worthwhile. I had made a difference in their experience and, hopefully, in their careers.

Cpl Diandra outlines how to achieve equivalence as a competent and respected Marine without compromising her femininity:

I don't like other female Marines very much. I like some, but most of them just flirt all day. They just want to talk and be all cozy with the guys, and they don't want to work. I guess it's easy to fall into that pattern because they're around good-looking guys all day and they want you because you're the only female in the shop. I don't want to be looked at as a female; I want to be looked at as a Marine. But those other girls just make it so hard because all they do is play the girl-boy games that I thought I could get away from in the Marine Corps. I don't like the flirting at work and the trying to be cute. But others try to be the opposite and try to be so butch. Every other word out of their mouths is the f-word and I think that's just disgusting. I don't know if they're trying to fit in or they just hang out with the guys too much and they take on that language from them, but I think it's horrible to see a pretty girl using language like that. They seem to think that to be a Marine, you have to give up being a woman, but you don't. You can be both. You can wear green all week and pink on the weekend. It's really not too big a leap; at least it isn't for me. When I'm off, I'm a wife. I cook and clean the house and do things that women do. Maybe that sounds stupid, but being a Marine makes me want to be a woman more. As much as I was a tomboy growing up, thinking that I wanted to be a boy instead of a girl, I am so happy now because I can play both parts and play them both well. I really, really love being a Marine. I think it's a unique experience that I could never get anywhere else in life. When I'm on the floor of my shop, the men treat me as an equal. In fact, I am better than most of them and they know and respect that. It's cool, too, because they know that if they have a question, they can come to me and I won't treat them like they're stupid. Instead, I'll teach them what I know, and if I don't know, we'll work through it together. If they see a woman in the office, though, there's an immediate disconnect because she's not doing what she should be doing, which is being a mechanic. They separate those females into a separate category than I'm in. It's obvious in their behavior and the way they talk. I have a bond with those guys that the girls in the office don't have and they'll never have until they're on the floor doing the job the Marine Corps gave us to do. I don't mean to minimize what they do because MIMMS is hugely important. What I'm saying is that there's a different level of camaraderie and respect that comes from being elbow to elbow with the guys, covered in grease and diesel. They definitely recognize me as a woman because when we're moving things that are heavy, they'll try to overcompensate for me if they think I'm not strong enough. I don't mind that; I think they'd do the same for a smaller guy that they respect. In that case, it's not a male-female thing. When they cross the line, though, I'll ask them to back off and let them know that I'm a big girl and can handle more than they think. I have pride, but I'm not going to be stupid and hurt myself just to prove a point. If I need help, I ask and they're always great when I do. They never, ever, act like I'm a weak link because they've seen what I can do and how much I care about all of us as a team.

Sgt Rocio describes how a female must often use a male's tactics when dealing with them:

I always have carried myself in a manner so people could not question my behavior. It's happened, though. There have been rumors. When it's happened, depending on the situation, I've either ignored it, or I've confronted the person who I believe was responsible for it. When I've confronted the offending Marines, they're shocked. I think they just can't believe that a female who, in their mind is "obviously" so screwed up, can accurately assess the situation and then have the balls to call them out. It's something that they've never, ever dealt with. In my experience, most are just bullies, used to getting their way with females and "weaker" men, so they take it as a given that they'll be "top dog" and that whatever bullshit opinion they hold will be regarded as "truth." Uhhh…no, sir, not in *my* Marine Corps.

All of the female Marines who self-identify as equivalent to male Marines clearly express that this status never compromises their femininity. In fact, all described being a Marine as simply another exciting part of themselves that they likely could never have developed as a civilian. In addition, they recognize that being a contributing member of the "gun club" garners them a sense of kinship, respect, and trust with their male peers that other females cannot achieve. By empathizing with the males and their perspective instead of condemning ignorant male behavior out-of-hand, these female Marines insist that they have stronger and more meaningful relationships with the guys based upon shared values and principles, those of United States Marine.

Sex Stigmas & Ramifications

In addition to Marilyn Frye's introduction and importance of the Willful Virgin to this discussion, she also discusses philosopher Robert Ehman's assertion that "sex itself" is value neutral, but insists that sexual intercourse does not occur in a social or psychological vacuum. She states that whatever acts people engage in as "sex," the parties to them are invariably moving and feeling within a distinctive, complex medium of power, myth, value, and the deliberate manipulation of desire for commercial purposes.[48] As evidenced throughout this work, all the respondents in this research recognize the very physical and sexual atmosphere of the Marine Corps. The females acknowledge this as a social fact,[49] but also recognize that they must manage their sexuality within it to succeed.[50]

The stories collected through the interviews codify a very definite judgment of female Marines who consciously choose to utilize their biological sexuality and socially constructed femininity to manipulate the environment, particularly the males within it, for their individual benefit. In these cases, awareness of female Marines' use of sexuality and the value-charged act of coitus, in this case a manipulative act, dominates the imaginations of Marines, male and female. One female Marine stated, "Before females entered the Marine Corps, I can't imagine what the guys talked about all the time. Now, it seems like it's nothing but our sex lives that interests them." No longer just a personal act, female sexual activity potentially undermines the good order and discipline of the military environment.

Although the female World War II veterans interviewed for this project speak of the rare "easy" girls in the unit, the real "problem" of rampant sexual activity between male and female Marines appears to have occurred within the All-Volunteer Force generation and continues today, even in Iraq and Afghanistan. With the gradual erosion of the separate spheres, males and females have had more and lax physical proximity to each other, and take advantage of that access. Males and females *want* to have sex, and they *do* have sex. Command problems occur when sex is used by females in ways that it cannot be used by males, to manipulate the system or individuals within it for one's own gratification or advancement.[51] One's sex inhibits opportunity or expands it.

The Deplorable Stigma of Weakness

Although making a point in a different yet related context, evolutionary psychologist David Buss writes, "As predators pick off slower and less agile prey, the remaining prey and their descendents evolve to be faster and more skilled at evading capture."[52] Its relevance here relates to the evolutionary adaptation that drives an individual to strive to be the faster and the more agile prey or, conversely, the faster or more skilled predator. In the Marine Corps, weakness simply, operationally cannot be tolerated. Weakness destroys the "fitness" of the unit, potentially reclassifying the unit as "prey" rather than "predator," a status held by the United States Marine Corps since its inception and lauded around the world throughout its history. Marines viewed as weak represent a liability rather than an asset.

In his myopia, Brian Mitchell writes, "To military women... emphasis on physical strength is anathema. When men in the military are encouraged to think that being strong and quick is good, the professional reputation of military women suffers. Because the services are committed to protect and advance women as equals, they devalue prowess as a professional virtue..."[53] It is undeniable that the physical demands of the military and warfighting are extreme, but the Marine Corps has consistently intensified its physical, and thus the leadership expectations, of its females rather than "feminize" the organization.[54] The two traits, prowess and virtue, are indeed tied explicitly together as professed by the Marine females who daily fight the good fight for equivalency.

Some females, eager to otherwise prove their worth in the organization, continue to be stymied by the stringent physical requirements of the Marine Corps. Captain Jordan, Naval Academy graduate describes the imperative of physical fitness, especially for females:

> Our battalion CO would take us on a run every Friday and he did run at a ballistic pace. They were tough and there was one time that I struggled. I was sick, but I didn't fall out but I did drop back. My company gunny told me, "Ma'am, you shouldn't have been out there," but I was a 2ndLt and that would have been the kiss of death if I had either not been there or had been there and dropped out. It just wasn't an option and I would prefer to have died on than run than to drop out and be labeled "weak." A female can

be the best officer in the world, but as soon as Marines see her struggle with either physical fitness or personal problems or being labeled, the odds increase exponentially against her up and down the chain of command. Physical weakness is the first thing that sets a Marine apart from his or her peers. It's much more dramatic for females, but it's true for males as well.

Captain Kami, too, is privy to the men's opinions of the "weak" females, but surmises that there may be a place in the Marine Corps for them, too, even to the detriment to female Marines as a whole:

The men bitch and complain with me around about the weaker women or the slutty women, so I never perceive their attitude to be "anti-female" in general. It really never impacted me because I think they felt comfortable enough around me to discuss these things and focus on an individual's behavior. There were only six females in my entire company at TBS, so it wasn't like there was a large population of women. Out of the six, including myself, the majority of us pulled their weight. There was one that was always "having cramps" or was on light duty or bed rest for one thing or another. She couldn't make it through the Endurance Course, and kept making excuses why she wasn't able to do one part or the other. I think she just wanted to eek through TBS because she knew that, once she was in the Fleet, the worst she'd have to do was the PFT. She was already assigned to be a Protocol Officer, so she just had to get through TBS to move on to bigger and better things. Her major in college was Public Affairs, and she didn't get that but she played her game to get what she wanted. So, she's not out there jumping over obstacles, calling for fire, or even leading Company PT in Okinawa, Diego Garcia, or Afghanistan. She's at dinner parties at the Pentagon schmoozing diplomats and I'm sure she does that very well. What's wrong though is that, instead of doing what was required of *all* Marine officers, she weaseled her way out of the very basics to live this high-profile life that grunts and every other Marine can never even touch, not that they'd want to but it's the principle. *That*'s what puts a bad taste in the mouths of male and female Marines alike, that unwillingness to really *live* the experience like everyone else does.

1stSgt Halona also speaks to the issue, but includes the effect of a female's weakness on the unit:

The men know a weak woman from a strong one, and they respect strength. But, if you can't pull yourself onto the bar or you can't stay with the pack for two miles, you've just made a name for yourself and it's not a good one. If you're constantly crying or whining over something, no one will respect you. Know why? Because we're *all* hot. We're *all* tired. We're *all* hungry. We *all* have blisters. We're *all* homesick. We're *all* scared. And, it goes on and on. You've got to get over this thinking that you're "special." Yeah, you might be important, meaning that you have something to contribute to the group or the mission, but you're *not* "special." You joined to be part of the team. That's what it's all about. It's *not* about *you*.

Sgt Justine, a corrections MP, describes the perception of weakness in women in relation to the big picture, especially in relation to the training conducted on the two Coasts:

[In California,] the men thought they were better than the women. A lot of that was fed, though, by the women who acted weak and acted like they couldn't do anything. There were males coming from San Diego who have never really seen female Marines before, and they are the worst. Then, you also have the males from Parris Island who saw us in recruit training and they saw the female Drill Instructors, especially at DI School, doing everything the men do. It was a different mentality, but when the San Diego Marines were leading the runs, their whole purpose was to create casualties of the females. You can sometimes gain their respect by keeping up and trying as hard as you can. You don't quit. You don't cry. You don't limp to BAS when you hear that there's a Company run Friday morning at 0630. If you act weak, you've screwed yourself and the rest of us [females] are screwed with you.

SSgt Tomasa describes a situation in which she refused to be "weak" regardless of the pain she endured and her anger at the women who succumbed:

One time, he [the Chief Cook] asked me, "Moore, how would you like some time off?" I was all over that, and he said, "just bring your 782 gear 'cause we're gonna have a little inspection." I was practically dancing to the chow hall, knowing I was gonna have the rest of the day off. Well, I got there and he said, "We're just gonna go on a little hump first." There was a convoy waiting to take us to the NTA [Northern Training Area]. The gunny was like,

"Moore, you better not fall out." I didn't want to disappoint him, but a couple of miles later the strap on my ALICE pack broke and it started rubbing to the point that I was bleeding. The CO walked past and said, "Hey, there, devil dog. You want to throw your pack up there in the truck?" But, the gunny was watching me and I said, "No, sir," and just kept on marching. All my girlfriends were riding past in the HUMVEE, and the gunny would say, "Yeah, look at those lazy chicks. You don't want to be with them." But, we were *lost*. We were humping and humping not even knowing where the hell we were going. The whole time, I was dying. After the hump, though, I had a few guys come up to me at different times, and they'd say, "Damn, Moore. I wanted so bad to fall out, but I'd look back there at you and there was no way I was gonna drop out. You just kept walking." I did fall back, but I never quit. I threw away that bloody t-shirt. There were a couple of other females that hung in there and finished. All the rest, though, were in the HUMVEE drinking water and laughing, and I was thinking, "Bitches!" [Laughs]

In the words of Nietzsche, "Independence is for the very few; it is a privilege of the strong."[55] This, in essence, is what females in the Marine Corps strive for: independence, empowerment, and success through personal strength and leadership, physically, mentally, and emotionally. Besides physical fitness, however, other obstacles exist that impede success.

The Stigma of Malingering[56]

Although alluded to in numerous other places in this work, two main manifestations of females' alleged malingering permeate this research: pregnancy and light duty, discussed here specifically in the context of stigmatization. Although being in a "light duty" status is not endemic to females by any means, their small numbers and propensity for "female" maladies magnify the prevalence of females on light duty, which creates stigma and its resulting ramifications.[57] Article 115 of the Uniform Code of Military Justice (UCMJ), Malingering, is extremely difficult to charge. For example, SSgt Nadine, a Career Planner, describes the situation witnessed in her ten years on active duty:

There are women who abuse the system. There are a lot who use the "I'm a female" so they get their way or try to get out of deployment. There are a few that *are* pregnant because they don't want to go to Iraq or Afghanistan. What they're not thinking about is the commitment that it takes to raise that child to adulthood. They're not thinking about that. They're afraid to go to war and they can't see beyond that fear. I really can't understand that kind of thinking. I mean, why are they even here? I've seen females who stay on light duty to get out of deployment, but when their re-enlistment came around and they had the balls to submit a package, they were denied. There's just too much at stake right now, and if you're afraid of doing the job you signed up to do which is "defend against all enemies, foreign and domestic," you need to go home and work at Walmart or wherever you might feel safe while the rest of us are here getting the job done. There are those few who squeak through, though. They're on light duty every day until about 90 days from when their re-enlistment paperwork is due or it's time to PCS. They'll try to bust ass to do well on the PFT, and they'll eek by because there's no paperwork to indicate that they shouldn't remain on active duty. They're just not being counseled properly and there's no documentation to support a malingering charge. There's no paper as they go from unit to unit where they display the same behaviors, and they just keep getting by. It's sad, but it does happen and it makes the rest of us look bad.

Sgt Ruby surmises how this self-imposed malingering evolves when a weak female enters the Marine Corps:

In my career, I've seen 18-year-olds come in and they're cute, and there are others that aren't so cute, but now, there're all these men just dying to be with them. Hey, we're *Marines*. We wouldn't be here if we didn't have something wild inside of us dying to break free. Anyone that would consider doing this isn't your mainstream kind of person. We're here because we're looking for more or different, so I think all of us have a rebellious personality to begin with.[58] We come in and do Boot Camp, but then, we have all these freaking *hot* male Marines around us, and it's on that individual female to decide how to handle herself. For some, it's easy to remain professional and keep the focus on her status as a Marine. For others, though, it's easy to fall back on what's easy, which is being a female who maybe isn't good at PT or who can't drill or who can't stand up to her peers [from a leadership perspective] and say, "Guess what? *I'm* in charge here." Those are the ones that fail. For those, it's easier to be on light duty all the time or marry some hard-up schmuck and start pumping out kids. I dunno the thought process that goes on, but I see it over and over again.

Cpl Erica returned from deployment, started a family, and now struggles against what she perceives as an undeserved malingering stigma:

One problem that I see that's the worst thing for females in the Marine Corps is women getting pregnant to get out of deployments. I got home from Iraq, got married, got pregnant, and was immediately tagged as a shitbird for getting knocked up. My response is that I'm *not* like these other women who get knocked up by some random guy to keep from going to Iraq. I went already and I'm slated to go to Afghanistan. I'm just a female, a career-focused Marine having a life in the middle of all this. There are the others, though, who've never been deployed. One of their big excuses after the birth is post-partum depression. I guess Andrea Yates really put the fear of God into the males about that one. I don't want to judge people, but if you're that psycho because you had a baby, you need to be given the boot now because you obviously cannot do your job as a Marine. On the flip side of that, though, I have also seen Marines come back from Iraq, who've seen some really serious stuff, all jacked up and wanting to re-enlist and maybe somebody should take a really close look at them before signing off. Like they can't be issued a weapon because of the medications they're on, they're not carrying their load and they're leaving it to the rest of us. We're short-handed enough as it is. We really don't need the sandbaggers dragging us down further. If you can't deploy, get off the tit.

The Practice of Manipulation & its Ramifications

In addition to malingering, CWO4 Renae also relates how females learn from a very young age to manipulate to get their way:

We did have the women perpetuating the stereotypes. They did what they could to get away with whatever they wanted. It happens with recruits today. I work with male coaches every day on the [rifle] range, and the recruits try to play the "I'm cute and you're a big, strong, wise man," and "I just don't understand this scary weapon," and batting the eyes thing, but our male Drill Instructors who are range coaches are trained to ignore it. At least a few in every platoon of female recruits try to do it. It is effective to teach the male coaches what they will expect with female recruits and what their expected response is. It starts early in the coach's training and we all watch each other to make sure no boundaries are crossed. It's funny, though, that in all these years I've spent on the ranges, I have *never* seen a male recruit try to schmooze a female Drill Instructor. I don't know why that is, but I've never seen or heard of that happening.

Women learn early, as children, to work their way with men. They cry or bat their eyes or otherwise manipulate to get what they want. These are behaviors females learn very early. If they're physically attractive, they might have different methods than those of girls who aren't as attractive. If a pretty girl pouts, it's just darling, isn't it? If an ugly girl pouts, it's just ugly and obnoxious. Who gets her way? The pretty girl, of course. Oh, but in our world with 93% males, pretty is less relevant and available sex works wonders.

Whether due to ambivalence or fear of allegations, too many male Marines adopt a "hands off" policy toward female Marines, often unwilling to intervene in "female issues" when in leadership positions charged with maintaining good order and discipline. The quality female Marine leaders, however, take it upon themselves to police those females who need it. In addition, a few like Captain Maya also educate the male Marines, arming them with information and the empowerment to address the problem in an appropriate, professional manner:

Some of my male officer friends have contacted me about these things. One male friend of mine who's an instructor called me on the phone. He was at the pool and he said, "I've got a female lieutenant who says she can't go to the pool because she's on her period. Is that true?" So, I told him to tell her to stick in a tampon and get into the pool. And, he goes, "I'm not telling her that!" So, that's what I teach now. I teach the instructors. I arm them with the right information. I always tell them, "Don't get me to come correct your women. *You* correct your women, but here's the information you need to give them. It's all about education. But, these guys, a lot of these instructors are infantry officers, and the last time they worked with women was at *their* TBS. So, it isn't comfortable for them. They don't mean to do bad things, but sometimes they say the most ridiculous stuff because they just don't know any better.

For female Marines, personal appearance and physical standards are different than for males. With the Marine Corps 93% male, it is obviously easy for that percentage of the force to be familiar with the standards that they, themselves, adhere to. The other 7% with different standards remain a challenging enigma.

Why Female Marines are Harder on Each Other

The data resonate with descriptions of female Marines expressing the imperative to be harder on other females because: 1) the female won't police herself, or 2) because the men won't or feel like they can't correct the female. In these unfortunate situations, strong females, conscious of the fact that the behavior of one bad female reflects negatively upon the reputation of all, step in to correct the problem. For example, Captain Jordan cites both pragmatic and philosophical aspects of the issue when identifying why female Marines are harder on other females:

> We had a female sergeant on the last deployment who was really hard on the females. The males never complain much, so I think it's not an issue with them. I think the female leadership tends to be harder on the females for two reasons. The first is because they want the females to succeed. They've set the example that it *can* be done and they expect those young females to tow the line. Conversely, they don't want those young females to screw up because it *is* a direct reflection on them. I think this is because we, as a society, put the onus of training on the nurturer who is typically the female. Just like a mother is responsible for her children, the female Marine is responsible for the behavior other females. Sorry if this sounds Freudian, but I think there is a lot of baggage we bring into the Marine Corps from our little homes and churches and communities. Some things are difficult to recognize and even harder to change once we recognize what's going on.

Sgt Charmaine recognizes the problem and deliberates about why female Marines perpetuate rumors and attempt to manipulate their environment:

> Female Marines are harder on each other than they are on male Marines. We have to be. From what I've seen, females are *not* the ones that start and perpetuate rumors, unless they're the weak women that have bad reputations and, therefore, nothing to really lose regardless of what is said. If they can create hate and discontent, they do it because it makes them "interesting" or, in

their own minds, "powerful." Some of these girls are just poison. What's really funny, though, is that these crappy women are usually only "interesting" to the low-ranking males or NCOs who are too stupid to know any better anyway and only want to get laid by these slutty girls. So, they get laid by the sub-standard males, so I guess it's all good. [Shrugs] A *quality* female Marine's reputation is going to stand, even if she does have to put up with stupid shit now and then.

Not always an easy task, GySgt Pamela relates why she has taken up the mission as a mentor and also as one who also potentially suffers from the bad behavior of female Marines:

Sometimes, I feel like whatever we do, we'll never be viewed as equal. We take one step forward and two steps back. Maybe not even of our own doing, but because we're all lumped together into this class of "women." The rumors and all....I have to admit that I was always harder on women than I ever was on men when I was a young NCO. I expected more out of women Marines than I ever would a man. I think because I wanted them to *be* better. I wanted them to be *viewed* as better than the marginal expectations, so I just pushed them harder. I didn't want to have them deal with, "Oh, *she* gets away with this" or "Oh, *she* gets away with that" because she's a woman.

Captain Gail bridges the gap between why females are harder on other females, and leads into a reason why females are not always the "best" solution to solve female "problems":

Women are definitely harder on other women in the Marine Corps. I think, too, that we are harder on ourselves individually because we carry this burden of having to prove that we're just as good as the males and to prove that we belong here. Individually, we set this higher standard for ourselves and it carries over to our concern for our collective reputation. If I'm busting my ass, I don't want LCpl Susie Rottencrotch making me look bad just because we're both female. I've been to units where there will be a senior female SNCO who brings all the female Marines around her and says, "Look. Let me school you on the ways of being a female Marine." Part of me thinks that's a good thing because it puts it all out there. It establishes the fact that women *can* make it in this gun club, and it builds camaraderie among females. There are some benefits to that. The young females know that there is someone they can go to with questions and issues that a male SNCO might not really have any experience with. At the same time, though, looking in from the outside, the men are saying, "Oh. There go the *females* again." They're wondering what that's about and probably feel left out. So, we bring it on ourselves when we segregate ourselves, even if it's with the best intentions. That allows the leadership to be split, which should never happen, but I've seen it more than once. If that male SNCO has a problem with his female corporal, he'll go to that female SNCO and tell her or ask her to square away that female. The only response is, "No. She's *your* Marine. *You* square her away." As an officer at my last duty station, Cherry Point, I was the adjutant and we had a female Marine who alleged that been raped in the barracks by three individuals. As the legal officer, on Saturday morning I got a call from the CO who asked/ ordered me to go make sure the female was okay and to check on her in her barracks room to see if she needed anything. As a good Marine officer, I was like, "Roger that, sir. I will go do that." The whole time I'm driving over there, and my husband said the same thing, I'm thinking, "Why am I doing this?" And the answer is, "Because I'm a female." I would like to think that he called me instead of her male OIC because he trusted me, he trusted my leadership, he trusted me to go into a delicate situation and deal with it appropriately. In my heart, though, I knew that I was being sent to deal with her and this delicate issue of rape instead of her appropriate chain of command because I am a woman. Her SNCO should have been there for her. She's a PFC who's been on the base for three weeks and I'm walking in there, an officer that she's never seen before in her life to talk to her. I think it was a very poor way to deal with the situation from a leadership standpoint.

Many NCOs, SNCOs, and officers concur that Marine leadership should be leadership of males as well as females. Male Marines must know the female regulations just as the female must know the regulations for males. True to the leadership traits and principles, Marine leaders of both sexes must be consistent and tactful in dealing with their troops. For those males who fear allegation, GySgt Claire offers advice, "Don't be afraid to be a Marine. What you do for your males, you do for your females. Your reputation will protect you from false allegations. Don't set yourself up for it, but don't be afraid of it either."

Female Marines recognize the varying levels of competency, commitment, and integrity exhibited by females within their ranks, but are frustrated by those who discredit the positive efforts of those who want to succeed. Many females struggle primarily against the stereotypes perpetuated by *other females*, not against male hostility or institutional misogyny toward them as individuals or as a cohort. However, they also struggle against male leadership when the men refuse to correct the females in their command out of ignorance, convenience, or fear of allegations.

Although highly structured and focused externally on national security, the immediate world of female Marines is one of innuendo, rumor, and stigma where flaunting sexuality or displaying weakness of mind or body is viewed in the harshest terms by everyone.[59] In this environment, female solidarity based upon sex/gender is simply not an option.[60] Each female must always be on her guard, anticipating that the actions of another will upset the delicate balance between males and females and, thus, bring discredit to the entire female population. Occasionally, female friendships are forged. However, the fundamental concern is that individual females possess the potential to destroy reputations and relationships that precede them. Successful females must be hyper-diligent to preserve the reputation they have worked so hard to achieve as equivalents with their male counterparts.

Endnotes

1   Kanter, "Effects of Proportions," 968.

2   Gerhard Kümmel, "When Boy Meets Girl: The 'Feminization' of the Military: An Introduction Also to be Read as a Postscript" *Current Sociology* 50, no. 5 doi: 10.1177/0011392102050005002 (September 2002): 617-618.

3   Ibid., 629.

4   Brownson, "Battle for Equivalency," 3.

5   King, "Women Warriors," 1.

6   ALMAR 046/12, Unclassified, R 271120Z NOV 12, Subj: Change to the Physical Fitness

Test (PFT), CMC Washington DC MCCDC C461TP/08AUG2008// and "Marines Delay Female Fitness Plan after Half Fail Pull-up Test," *New York Daily News,* US, January 3, 2014, http://www.nydailynews.com/news/national/marines-delay-female-fitness-plan-fail-pull-up-test-article-1.1565216.

7   Kanter, "Effects of Proportions," 972.

8   Carol Cohn, "'How Can She Claim Equal Rights When She Doesn't Have to Do as Many Push-Ups as I Do?': The Framing of Men's Opposition to Women's Equality in the Military" *Men and Masculinities* DOI: 10.1177/1097184X00003002001 (October 2000): 135-139.

9   Brownson, "Rejecting Patriarchy," 4.

10 Brownson, "Battle for Equivalency," 12-16.

11 There are too many to list; referenced in this work, specifically, are Browne, *Co-Ed Combat*; Maginnis, *Deadly Consequences*; Mitchell, *Women in the Military*; and van Creveld, *Men, Women, and War*.

12 Dee Brown, *The Gentle Tamers: Women of the Old Wild West* (Lincoln: University of Nebraska Press, 1958), 127.

13 Titunik, "Myth of the Macho Military,"143.

14 Brownson, "Rejecting Patriarchy," 3-4.

15 Kanter, "Effects of Proportions," 968.

16 Culver, "Women in the Service," 67.

17 Laura Sjoberg, *Gender, War & Conflict* (Malden, MA: 2014): 170.

18 Titunik, "Myth of the Macho Military," 139.

19 Meštrović, *Postemotional Society*, 148.

20 Kanter, "Effects of Proportions," 968.

21 Brownson, "Rejecting Patriarchy," 5.

22 Brownson, "Battle for Equivalency," 21.

23 Brain, "Sex, Incest, and Death," 192.

24 Brownson, "The Battle for Equivalency," 17.

25 Brownson, "Rejecting Patriarchy," 5.

26 Culver, "Women in the Service," 66-67.

27 Kanter, "Effects of Proportions," 982-984.

28 Williams, *Gender Differences*, 86.

29 Culver, "Women in the Service," 66.

30 Brownson, "Battle for Equivalency." 17.

31 Buss, *Evolution of Desire*, 58, 66-67.

32 Ibid., 268-269.

33 Johannes Han-Yin Chang, "Mead's Theory of Emergence as a Framework for Multilevel Sociological Inquiry," *Symbolic Interaction* 27 (3) (Summer 2004): 405.

34 Brownson, "Battle for Equivalency," 20.

35 Ibid., 66-70.

36 Blau, *Exchange and Power*, 80.

37 Ibid., 14.

38 Buss, *Evolution of Desire,* 3.

39 Brownson, "Battle for Equivalency," 5-6.

40 Buss, *The Evolution of Desire,* 74.

41 Kanter, "Effects of Proportions," 973.

42 Ridley, *Red Queen*, 132.

43 Jesse Lee, "The President Signs Repeal of 'Don't Ask Don't Tell: Out of Many, We are One" *The White House Blog* December 22, 2010, http://www.whitehouse.gov/blog/2010/12/22/president-signs-repeal-dont-ask-dont-tell-out-many-we-are-one.

44 Kanter, "Effects of Proportions," 968.

45 Brownson, "Rejecting Patriarchy," 5.

46 Brain, "Sex, Incest, and Death," 194.

47 Jennifer Boldry and Wendy Wood, "Gender Stereotypes and the Evaluation of Men and Women in Military Training" *Journal of Social Issues* 57 no. 4 (Winter 2001): 701-702.

48 Frey, *Willful Virgin*," 41.

49 Emile Durkheim, "Sociology and Social Facts," in *Social Theory: The Multicultural & Classic Readings*, ed. Charles Lemert (Boulder, CO: Westview Press, 1993), 81.

50 Brownson, "Battle for Equivalency," 16-17.

51 Buss, *Evolution of Desire,* 97.

52 Buss, *The Murderer Next Door*, 43.

53 Mitchell, *Women in the Military*, 145.

54 Brownson, "The Battle for Equivalency," 21.

55 Walter Kaufmann, trans., *Beyond Good and Evil: Prelude to a Philosophy of the Future* (New York: Vintage Books, 1966): 29.

56 Uniform Code of Military Justice (UCMJ) Article 115. Malingering: Any person subject to this chapter who for the purpose of avoiding work, duty, or

service (1) feigns illness, physical disablement, mental lapse or derangement; or (2) intentionally inflicts self-injury; shall be punished as a court-martial may direct.

57 Lynn Zimmer, "Tokenism and Women in the Workplace: The Limits of Gender-Neutral Theory" *Social Problems* 35 no. 1 (February 1988): 65-66.

58 Ricks, *Making the Corps*, 23.

59 Brownson, "Battle for Equivalency," 15.

60 Titunik, "Myth of the Macho Military," 143.

Chapter Seven

The Balance of Power: Authority vs. Victimization

Non-veteran civilians imagine that the military exerts excruciating expectations and control over its members.[1] The fact, however, is that Marines experience great latitude to exercise personal freedom and make decisions within the military structure and are encouraged to do so. It is the essence of Marine Corps leadership, trust, and confidence. From initial training at Boot Camp or OCS, Marines are taught to take active leadership roles and to exercise good judgment when making decisions. This begins at the squad level. Accountability is paramount to successful leadership with the leadership traits and principles guiding appropriate decision-making.[2]

Unity of effort, legitimacy, and perseverance are essential to success in today's "small war" environment.[3] In this environment, no man or woman is all-powerful. Brian Mitchell accurately identifies the situation in the Marine Corps in which leaders exert "authority" through competent leadership rather than "power" through coercion, stating:

> In any civilized military, the force exerted by superiors over subordinates is not power; it is authority. Men who exercise authority acknowledge that they themselves are subordinate to others. Men who wield power answer to no one. In the armed forces of a democratic republic, the only power that should matter is firepower.[4]

Captain Kami, a Military Police Officer, offers an informed and insightful opinion regarding leadership and individual behavior:

> Abu Ghraib happened right before I got to Iraq. I think Abu Ghraib and other atrocities committed in Iraq are perfect examples of small unit leadership just not being there or gone horribly wrong…I personally believe, as an MP officer and as an officer

now supervising Drill Instructors, that human beings given complete authority and control over other human beings will *enjoy* dominating and humiliating and expressing that control over other human beings in their charge. There has to be checks and balances in place to ensure that abuses don't occur. Otherwise, "bad" things *will* happen and that's why the military has a structure, a staff, dedicated to supervision to ensure that "bad" things don't happen. I hate to sound negative, but subjugation, I think, is human nature. Maybe, not the "best" of human nature, but I think it's dangerous to ignore what's *really* going on versus what we *hope* is going on.

The distinction between the concepts of "authority," "power," and "victim" related to sex/gender relations is critical to this discussion. Evolutionary psychology and sociology tenets assert that females are active agents in social and sexual relationships with males and that these relationships are negotiated, a process of give-and-take.[5] Thus, the complex interplay of human sexuality, mating behaviors, and social control in the military demands pragmatic investigation.

Evolutionary psychologist David Buss acknowledges this intersection asserting that our evolved strategies of mating are highly sensitive to legal and cultural patterns.[6] Similarly, the biologically instructed mating strategies of the two sexes remain the most important aspect of Buss' theory to the current research. Crucial to this investigation, *biological* imperatives correlate with *social* dictums -- not in any direct, necessarily causal way, but as a way to pre-structure or at least put pressure on free decision-making. Controversial, titillating, and admittedly by Buss "not nice," scientific research findings about human sexuality and mating identify the very foundations of human motivation and behavior.[7] Similarly, recognizing that human society consists of a variety of exchange relationships between and among individuals, sociologist Peter Blau advances the concept that social attraction is the force that induces all human beings to establish social associations on their own initiative.[8] On the most basic social level, one individual may possess and/or command a commodity or service desired by another individual. Anyone who commands the sought-after commodity or service attains power over any other who desires it by withholding the desired item contingent upon satisfaction of his or her need. Sharing or relinquishing the desired item remains with the possessor.[9] The desired item in social relationships among military service members is sex. Another is trust. The tensions created between them compromise relationships on the micro- as well as macro-level.

Within the biological and social interaction parameters described previously, per Buss and Blau, a female Marines' sexuality, beyond her own control, is commodified. In traditional cultures, rather than the woman herself maintaining power over her sexual commodity, if she is unmated, it belongs to her male kin for protection until it is transferred to another approved male.[10] It follows that, in theory, in contemporary American society, if a woman could, within her own capacities, wholly manage her physical and sexual capacities against encroaching males, the need for male "protection" of her sex would not be necessary. The military structure endeavors, in a non-gendered manner, to ensure that no solitary person within the chain of command wields unlimited power over another person.[11] Feminist philosopher Camille Paglia also insists that women's association with sex should be seen as her source of greatest power, not as the root of her oppression and abuse.[12] Why, then, do sexual harassment, assault, and rape in the U.S. military continue to capture headlines?

The Evolution of Sexual Harassment in the Marine Corps

As previously stated, in response to the 1991 Tailhook scandal, the Secretary of the Navy announced a "zero tolerance" policy on sexual assault by members of the naval services, which includes the Marine Corps.[13] Katzenstein and Reppy's edited work with that very title from 1999 provides little value to this discourse a mere 16 years later simply because their "military culture" through civilian intervention has evolved substantially since its writing.[14] It is true, however, that at the point in time when female Marines served in capacities most equivalent to their male counterparts, their categorization as victims of sexual harassment, abuse, assault, and rape eclipsed their many successes. The question still unanswered is: How can a female be a competent military leader if her sex potentially renders her physically vulnerable and professionally impotent within her own organizational structure?[15]

One key assumption from an evolutionary psychology perspective as stated in the Introduction is that females as well as males possess a fundamental interest in successfully reproducing, with sexual interest and desire being a motivating force in humans' lives. We operate as sexual beings throughout our lives. The strategies in mating, however, differ for each sex.[16] Historically, as well as in contemporary discussions of the issue of sexual harassment, the radical feminist position exclusively dominates. They allege that sexual harassment is intrinsically an exertion of a male's physical superiority and power over that of a female, with the male using her sex to manipulate, degrade, or physically possess her.[17] The problem with this position is that it assumes that females are not active agents in universally performed mating rituals[18] or day to day conditional interactions through which actor and environment respond to one another through mutual conditioning, which affects community construction or disruption.[19] Further, it establishes females as "victims" in their own minds and in the minds of predatory males. Adopting this position also assumes that re-conditioning the males can "fix" the problem; it places the onus for inappropriate sexual behavior wholly and unfairly on the males. Hence, since the early 1990s, and especially in the wake of Tailhook, the Marine Corps provides professional development instruction on sexual harassment, abuse, assault, and rape. The Corps added Equal Opportunity, Victim Witness Liaison Officers (VWLO), and Victim and Witness Assistance Coordinators (VWAC) billets. Marine Corps Orders have been written and published, and databases and hotlines established. These will not mediate the issue of female sexual vulnerability, however, until the military organization and civilian society acknowledge and address the simple fact that females and males are inherently biologically complementary, but disparate beings who interactively negotiate sexual relations.[20] Thus, the onus is on *both* actors to engage appropriately in interactive sex behaviors. [Please note that this does *not* include forcible rape, which is a heinous violent crime, in which an individual is ambushed by a predator with no opportunity to enact personal agency to deflect the attack.]

In our contemporary American postemotional society, we savor relaxed norms and fewer restrictions on our behavior through tradition, community scrutiny, etc. on social interactions and mating rituals between females with males. The result, however, is greater opportunity for sexual advances and even their deflection to be misaligned with intent and misinterpreted by targets and observers. Systemic of postemotional American society, in general, Meštrović identifies this quandary:

> Historically, people did not get as indignant over as many things as so many contemporaries do…It is quite another matter when people today feel victimized by a look, word, phrase, or touch based on the subjective interpretation of the alleged victimizer's motive and on the synthetic construction of their victimhood status based on membership in a specific group deemed to be a victim. The resulting indignation by victims is often manipulative, resulting in a new form of resentment.[21]

Although a small sub-sample in this research, female Marines of the Free a Man to Fight generation talk about males "being fresh" with them, but none expressed feeling intimidated or disrespected, nor did they witness any overly-aggressive behavior towards their female comrades. They lived in separate spheres. Even their social worlds were separately maintained and supervised. In the 1940s for the Free a Man to Fight generation, sexual and social roles were rigid.

Miscommunication or, perhaps, misrepresentation of intent between the sexes, however, remains temporally consistent. In the AVF era, Women Marines expected male Marines to do stupid things to get their attention and gain their favor. Emerging from external societal [read: feminist] expectations and demands, the AVF was an experiment plunging females in to a high-stakes, male-dominated petri dish, *qua* social and occupational military environment of negotiation and compromise, with no playbook as guidance. Zimmer is largely correct in her criticism of Kanter's "tokens" as embodied by AVF female Marines. Women's occupational challenges cannot be alleviated by achieving numerical equality.[22] Instead, attention and investigation must be focused on socially constructed sex-based attitudes and biologically motivated behaviors.

As with the WWII experience of females, no data obtained through this research reflect female failure or fear resulting from the openly hostile environment of that era. Acceptable interactions between men and women, sexually, socially, and occupationally, were admittedly in flux.[23] Post-Tailhook (1991) female Marines, however, specifically describe an environment of confusion about "appropriate" behavior between the sexes and the ensuing resentment of *both* sexes when the ambiguous lines are crossed. Again, the question remains: How can a female be a competent military leader if her sex potentially renders her physically vulnerable and professionally impotent within her own organizational structure? Ultimately, if not herself, who is responsible for her empowerment and subsequent career?

Major Lauren describes her early days of a 30-year Marine Corps career beginning in the 1970s as not being focused on being a woman or a man, but just being buddies, hanging out and having a good time. She describes the environment, a gun squadron, into which she walked as the first female in MOS 6025, Harrier Engine Mechanic:

> When I first got there, I was not welcomed. The word "cunt" was used to describe me. Here I was, a female coming into a totally male environment. It was a tactical squadron. They'd never had a female around, *ever*. There were no female clerks, no females, period. They viewed my presence as an all-out invasion of their masculine domain and made it clear that they weren't happy about it. Back then, of course, the guys had Playboy pinups all over the place. So, there are naked women staring at me from the bulkheads. It didn't bother me, but I had that bit of fire in me, so I hung up Playgirl pinups in my area. As soon as I put my naked guys up, all of the other naked pinups came down. I can only guess that they were pretty shocked by the pictures, one, and the fact that I had the balls to hang them up, two. [Laughs] No one ever even said anything about it. It just happened that way. So, I took my naked guys' pictures down and we continued to march. Could I have gone to someone and said, "Hey, I'm really offended by this?" Sure. I could have, but instead I used their own tactics and they didn't like it.

Lou Ann further describes an incident of harassment and her and her unit's response:

> I spent a lot of time being squared away and looking sharp. One of the guys had taken a dump in my cover. This is one of my favorite stories that I tell my females today. "You think *you're* being harassed?" Listen to some of *my* stories. So, here I am at work and I have crap in my cover. I went to my gunny and I said, "I really don't know what to do with this situation." Nothing could

have prepared me for something like that! He found out who did it. Back in the day, when we could haze and harass, that boy got hazed and harassed to the point that, I'm sure, he thinks about me now every time he takes a dump. All the other males thought he was an idiot, and everybody knew about it and got their licks into him. The issue wasn't that I was a female. The issue was that civilized people just don't *do* that stuff, especially not to fellow Marines. We didn't have "sexual harassment" or "sexual assault" like we do now. I just used my chain of command and that was fully effective because Marines policed themselves to a large degree. This was just one idiot who chose to constantly screw up. My Marines took care of it.

MGySgt Sally also talks about the relatively early days of the AVF, affirming that there was no language for sexual harassment but plenty of inappropriate male behavior:

On the harassment issue, I would have said I was never harassed in the sense that one would think of harassment then but, looking back as an adult and I know what harassment is now, there was all kinds of that. For example, I remember as a young cook watching the sergeant in charge of my group carve a scallop into what looked like a vagina. He showed it around, and ha ha ha. Too funny, right? It was no big deal, and if I'd acted silly about it, the guys would have known they'd gotten to me. I think I went into it [the Marine Corps] knowing that stuff like that was gonna go on and I just rolled with it. Like, I wasn't going to let that offend me. It wasn't directed at me; it was just stupid boys acting stupid.

GySgt Claire relates a story of being sexually harassed, in the absence of language to describe it, as young female Marine in Okinawa in the late 80s:

In Okinawa, [in the chow hall] we wore burgundy smocks and I always wore burgundy lipstick to match my smock. I worked for a really obnoxious mess chief. He would always say, "Smith! Girl, I *love* them lips." He was just a dirty old man. He'd stand in his office window, watching the girls come through the chow line and he'd say, "Who's that? Hook me up with her!" No one ever said anything to him about his behavior. He was brash and loud. He always did whatever he wanted to. He started making me uncomfortable. It didn't bother me that he talked to me like that 'cause I could hold my own. What bothered me was that everybody else heard him talk to me like that. They had started asking, "What's up with you and the Master Guns?" That was when I had to do something. I stopped wearing that lipstick and started wearing a clear gloss instead. Finally, I guess he realized this and he said, "What happened? Girl, you know I love them lips. You better put that lipstick back on." I told him, "Master Guns, I don't like you saying stuff like that to me." He said, "What? You don't like compliments?" I explained that it wasn't a compliment to me when you're talking about my lips and the things you want to do to them. That's *not* a compliment. He said, "Girl, you know I'm just playing. You shouldn't have no sexy lips like that." He just brushed it off because that was just his nature. So, I just wore my lip gloss and ignored him when he got obnoxious. I started thinking about it, though, and if there was a female working there with him and she wasn't as together as I was, I could see how that would make her not want to go to work. I could see how she would feel demeaned and intimidated. What made it even crazier, though, is that he talked about women like he could do whatever he wanted to with them, but he wasn't attractive at all. I mean, *not even*. Today, that's totally sexual harassment, but back then we really didn't have the language to describe it or the structure to deal with it. It was just what women had to put up with and manage the best we could.

CWO3 Phoebe professes her disgust at the endemic sexual harassment in the AVF generation, but also speaks to the imperative that a female, herself, must mediate the situation by not infusing her sexuality into her worklife:

I was always so offended by the label "WM." We weren't "Marines." We were only fooling ourselves if we thought we were. We ran a sub-standard PFT. We couldn't deploy or even go to the field unless it was in numbers to justify billeting which was, pretty much, never. We were novices with the rifle. We weren't allowed to touch heavy weapons. We were just "WMs." It should have been WD, meaning "Window Dressing," and that was made clear to us whenever a male felt like it should be. At that time, we lived with that and all the little acronym ditties that went along with it. What could we say in response? Nothing! It was *true*. We weren't allowed to be any more than what they dictated us to be. I am a proponent of females doing what they need to do to survive and succeed in this environment. Yes, you can still look really cute in a pair of jeans with your hair down, but when it's time to be a Marine, all that civilian femininity shit gets put on a shelf and you're in *Marine* mode. You can't be both, a cute little civilian chick and a Marine, at the same time. It doesn't work, and don't confuse the males with it. They simply can't handle it.

Captain Sophia, a recruiting officer, describes an event that occurred when arriving at her new duty station in San Antonio, Texas:

When I arrived, I got a call from one of my local recruiters. I answered the phone, "Captain Sophia," and he was, like, "Wow, ma'am, if I had known you were gonna answer the phone like that, I would have thought I was calling a different kind of establishment." I had never met this young man before. *Never met him.* He's a sergeant. I'm a captain. I'd just gotten here. I'd been here, like, a week, maybe, and I was like: "Oh. My. *God!*" I was flabbergasted because no one has *ever* talked to me like that. So, I said, "Excuse me. Who *is* this?" And, he said who he was, and I said, "I think we're done with this conversation," and I hung up the phone. It wasn't intentional harassment; it was just popping off, but that's still unacceptable. So, I immediately told my Master Sergeant to square him away. He had that young man up here in about two-and-a-half heartbeats and gave him a new meaning of life out on the porch, and then he understood. Poor kid…he didn't talk to me for, like, two months 'cause he was terrified of saying the wrong thing again. We're fine now, but I imagine that he *thinks* now before opening his mouth.

MSgt Madeline offers her perspective on the current situation with 20+ years of experience to back it up:

I just don't see women in the Marine Corps as being victimized. I haven't seen sexual harassment as an issue undermining good order and discipline. I have seen one rape case in my career, but it was a *civilian* guy who broke into a female Marine's apartment. I see the horror stories on the news, but I just haven't seen it in reality. Recently, I saw something on CNN about women in the military having eating disorders because of the pressures to be thin and physically fit, and that the numbers were high. I don't know how much they focused on Marines, but I was really surprised. I mean, I'm a Master Sergeant, and I've seen a lot of women come and go but, in my experience, I've never seen a single female Marine with an eating disorder. And, no one can say, "Oh, that's just because it wasn't 'identified' as an eating disorder." Bullshit! In the Marine Corps, to do all that we do, it takes eating to be fit and to *stay* fit. If you're not eating, you're passing out. If you're not eating, you're not successful at PT. "Taking care of Marines" means being aware of every aspect of their performance, and it's our duty as SNCOs and officers to ensure that things like this don't get out of hand. The American media focuses on the negative because it draws viewers. I'm not saying that bad things don't happen, but I've never experienced it in over 20 years of service in a wide variety of places.

GySgt Hailey, currently an EO Advisor, describes the situation today:

I hear young Marines say silly sexist things without even realizing what they're saying. This is because we come from a sexist society where there is language about females in extremely sexy ways such as in advertising, music, and video games. So, I think they say silly, stupid, immature things not even comprehending what meanings might be construed from their words. I rarely see or hear any malicious intent, though. Mostly, they don't mean any harm; what runs through their brain rolls off their tongue. The young ones…well, you know what they say…They're "young, dumb, and full of cum." Sorry for being crude, but that will never change no matter how many EO classes we subject them to.

MGySgt Elizabeth recognizes the importance of the situation, but offers a bit of practical advice:

These females shouldn't get all up in arms because, even in today's Marine Corps, you can't be making allegations every time something mildly sexist is said. Some things are innocent or ignorant. Some *are* malicious. I think we've lost sight of those categories in society and that's bled over to the Marine Corps. Instead of worrying about getting our feelings hurt by everything, we need to focus on the intent of what's actually being said or done. In my experience, most of what's said and done is innocent and it usually takes calling attention to it for the behavior to cease. It usually doesn't take much, just the strength and integrity to stand up for yourself.

Speaking specifically to gender relations and innocent vs. malicious behaviors in the military, Meštrović further states, "…whereas traditional societies devised ways of deflecting the male gaze…contemporary societies have established a new, double-bind puritanism in which the body is systematically displayed…at the same time that the gaze is just as systematically prohibited and often punished… it is really quite new that modesty in Western culture is decreasing while the policing of desire and thought is increasing."[24] This insight is of utmost importance to on-going sex/gender relations, current law, and proposed legislation relevant to the U.S. military. The data acknowledge an isolated, sex-skewed, hyper-physical, and youth-centric Marine Corps experience for females. This is compounded because, at the same time American females flaunt sexual freedoms they have achieved in American society, the female sex continues to be commodified by the mass media to titillate audiences and by marketing corporations to sell products. In this environment of mixed messages, abandonment of restrictive expectations for "proper" behavior of and between the sexes, and the human desire to mate and procreate, young male and female Marines find themselves in a quandary. Biological and social behaviors, rather than maintained in separate, corresponding spheres are now exposed, juxtaposed, and hypermanaged, often in highly artificial and contradictory ways.

Personal Responsibility vs. Leadership Failure

The data confirm that if male Marines are not actively malicious in the things they say and in the behaviors they enact toward females, they are likely simply going through innate mating behaviors. Due to inexperience, these sexual advances may manifest as miscommunicated or subject to "misinterpretation" by the recipient. David Buss notes for example that, although stalking is now illegal in all states in America and in most countries throughout Europe, research confirms that it's a surprisingly common strategy of human mating."[25] The data gleaned from female Marines across time and throughout the rank structure state very clearly that many alleged female "victims" knowingly set themselves up in situations of misinterpretation or outright manipulation. Often, they want too badly to be part of "the group." By engaging in recreational short-term mating behaviors,[26] they simply fail to establish appropriate boundaries with their male counterparts.[27] Some become victims of jealous and sometimes murderous rage.[28] Although in no way condoning "blaming the victim," the female Marines interviewed live in an environment in which management of one's self and one's sexuality is critical to success in their social and professional lives.

In true Marine Corps spirit, MCO 1752.5 of 28 Sep 2004 states, "Leadership is the key to sexual assault prevention…"[29] However, what female Marines express, supporting the evolutionary psychology and postemotional ways of thinking, is that *individuals*, both female and male, regardless of the quality of their leadership often make bad decisions that result in sexual harassment, assault, and even rape. They also make it clear that "legitimate" sexual assault and rape are criminal offenses best dealt with under the UCMJ and *not* as Equal Employment Opportunity challenges. Unfortunately, though, real life scenarios are never as clear-cut as they may seem in theory in Marine Corps Orders and related legislation. "Leadership" is left to clean up the mess after the fact.

MGySgt Geneva, whose career spans the AVF and post-Tailhook generations, describes the current state of EEO as a Pandora's Box:

> I tell my staff sometimes that my life has changed so much just in the past few years. It's not just a matter of being a higher rank; it's the change in the atmosphere. I tell my staff to be cautious in how they talk to those young Marines. You have to be real careful because those young Marines can no longer be treated the way we were when we were sergeants and below. You can't use profanity. You can't single anyone out in any way because it can be *viewed* as either preferential or derogatory and both aren't acceptable. It's changed a lot from the 80s to today. Yes, maybe Pandora's Box was opened with all of the sexual harassment legislation. Our focus in the Marine Corps has always been on taking care of our Marines. Now, though, we have to be *respectful* of them regardless of whether they've proven themselves or not. These are two very different concepts and one doesn't necessarily translate equally to the other.

Similarly, 1stSgt Bonnie describes the change in climate from the 1980s to today:

> In the 80s, we didn't have the language "sexual harassment." It just wasn't an issue. If a female had a problem with a male being too aggressive and she felt like she couldn't handle it, she took it to her chain of command. True, sometimes she was blown off. Usually, in that case, her buddies took care of it. Now, the Marine

Corps has become more civilized. It's not because of the women. We've always been the few and the proud. If we could keep to the same traditions, these "modern problems" wouldn't even be an issue and if they became an issue, they'd be dealt with accordingly. Drug use not tolerated; you're gone. Fraternization not tolerated; you're gone. Instead, now, we accommodate everyone's opinions. On the one hand, you have Marines now "interpreting" everything, every look, every move, every word. On the other hand, you have the opposite extreme, which is MCO 1752.5 of 28 Sep 2004, which is a by-the-numbers outline of how to handle sexual assault! It even has a flow chart! In generations past, the Marine who rapes would have been charged with Article 120 of the UCMJ, given brig time and a BCD [Bad Conduct Discharge], and sent packing. What we have done is gotten away from "right" and "wrong" behavior into something situationally relevant or negotiable. It's crazy!

In the present environment, 1stSgt Katrina speaks of a recent questionable "harassment" allegation at Camp Lejeune:

I haven't had any experience personally with sexual harassment. I have been a SNCO for many years and I just haven't seen it as an issue. So much of a woman's treatment in the Marine Corps is directly related to how she conducts herself. Yeah, there are sickos out there who stalk and men try to seduce in obnoxious ways, but a lot of dealing with that just has to do with maintaining your professionalism. I recently had a young female who was being administratively separated come to me who, while she was in school at Courthouse Bay, was supposedly harassed because the instructor grabbed her ass or something like that. What I *saw* was schoolgirl drama. She claimed that she now had psychological problems working around men, blah, blah, blah. It's interesting though, that she has "problems with men" but, while she was awaiting discharge, she acted like a slut, talked like a slut, and always dressed provocatively, barely within regulations, always wanting to know when the next Bosses' Night was. But, of course, that's why DoD employs the legion of counselors to help her restore her "quality of life," right? *She*'s the reason females have a bad reputation in the Marine Corps and I was glad to see her go, knowing full well, though, that there will be more to take her place in perpetuating the negative stereotypes. It's the few like her that make the males unwilling to trust us and proves to our critics that males and females just can't play nicely together without the situation becoming something perverse and hostile. It's just a damn shame.

Potentially serious legal issues as well as leadership challenges for Marines, alleged sexual harassment, assault, and rape must be treated as "real" events, often with serious consequences for all involved, until validated or proven otherwise. In some cases, however, the alleged sexual assault simply did not occur. CWO3 Leona describes a situation she recently experienced:

> I had a female student call me one night when I was the instructor on duty at Cherry Point. Somehow, I knew…maybe women's intuition…but I knew what she was going to tell me. She started hinting around that she had been out with a bunch of the guys. Just her and the guys. And, I stopped her and cautioned her. I said, "I'm not saying I'm not going to believe what you're about to tell me, but you need to be very careful about what you're going to tell me." She basically accused three or four of the guys, while they were drinking in a BEQ room, of gang raping her. I said, "Hey, okay, I believe you. But, now that you've imparted this to me, I have to follow through and take some steps." I warned her before she even told me that she had to be very sure that this had happened because disciplinary actions are extreme. Come to find out that, after eight or nine months of intense and ugly investigation, she had lied. So, here is a female who ruined these Marines' lives for nine months and, now, they have this stigma to carry with them. She ended up getting kicked out. What had actually happened that night was that she was late for a formation the next morning and concocted the rape story to cover her ass. I was like, was it *worth* it? She couldn't just take the damn Page 11 and learn from it? She was a PFC! No big deal. Instead, though, she ruined those guys' lives and careers, she destroyed unit cohesion, and she made the males not trust females, in *general*, just to cover her own sorry ass! That's not what Marines are about, and it sickens me that so many screw it up for all of us.

The data suggest that some of the females themselves are guilty of compounding the confusion and resentment surrounding appropriate and inappropriate social and sexual behavior in the Marine Corps. GySgt Marjorie shares her experience and the advice she offers her young Marines, both male and female:

I think the reason that I never experienced the "typical" female problems is because I never let sex be an issue. My presence here is as a Marine, and I will accept no less from anyone else. At my first duty station, I did some trash talking with the guys, but I also made it clear that we could be buddies but it would not progress beyond that. My private life, I keep private. When I'm in uniform, it's all about the Marine Corps and my fellow Marines. I think that's one of the mistakes our young female Marines make, trying to be part of that male group to the point that they don't establish the personal boundaries to protect themselves from misunderstanding and they wind up making themselves vulnerable.

First Sergeant Hillary adamantly asserts that the Marine Corps is not unique in the sociosexual experience of its females and should not be targeted as a hostile environment simply due to sex-skewed demographics:

Rape is rape whether it occurs at Camp Lejeune or Louisiana State University or wherever. It's an act of force and aggression, a criminal act. Are the frats on college campuses labeled misogynist because they don't even allow women into their pretty boy social clubs? Is the American Medical Association misogynist or racist or homophobic because most American physicians are predominately still white heterosexual males? Is the Marine Corps misogynist because it expects women to prove their physical and mental worth in the finest warfighting force on Earth? Honestly, I dunno, but our expectations are realistic and attainable. I would seriously warn critics voicing the whole "hostile environment" accusation to fact-check better. The concept of "rape" is in a really weird place right now in terms of definition, so don't say it's the military's problem. There are too many outstanding, well-adjusted women that do very well in the Corps. They're physically fit, occupationally successful, and respected up and down the chain of command. With 7% of the entire force female, if there is, in fact, one *bona fide* rape, however tragic, do not condemn the remaining 93%, our males, and the organization itself. I would caution the uninitiated not to lose sight of the thriving forest because of one ugly, dead tree that sticks out in the snapshot.

LtCol Abigail, a Marine Corps attorney, describes the relation between leadership and personal responsibility:

Sexual harassment and sexual assault are leadership issues in the

Marine Corps. I say that with the express understanding that *individuals* will do what *individuals* will be inclined to do. The Marine Corps environment, *de facto* and *de jure*, does *not* tolerate sexual harassment or any intimidating attitudes or behaviors towards females. Therefore, if a Marine assaults another Marine, the investigative committee better look extremely hard at the circumstances before jumping to the conclusion that the Marine Corps somehow fosters or condones criminal behavior. It's simply not true, and I grow weary of hearing about it in the media. If and when an incident does occur due to an *individual's* behavior, we have the chain of command, the leadership, in place to deal with it. We have a strong tradition of service to our country, our Corps, and to each other. We work very hard on educating each other and assisting each other with some of the issues we struggle with daily as Marine Corps leaders. Each Marine, male and female, should take that to heart. If you didn't care, you wouldn't be here. If you're here and you don't believe in and embody our principles, you need to be finding your way out.

As introduced in Chapter 4, the Rape, Abuse & Incest National Network (RAINN) rejects rape as a systemic problem. Rather, RAINN acknowledges rape as a conscious decision made by a small percentage of individuals within the community to commit violent crime.[30] Like RAINN, all female Marines interviewed who discussed this issue testify that a female Marine's personal conviction, chain of command, and available victim services sufficiently prevent sexual coercion. Violent sexual crime is a separate issue and the two should not be lumped together. Marine Corp leadership and legislation via Marine Corps Orders diligently attempt to provide appropriate and necessary expectations and training to ensure that all Marines know what is and isn't appropriate behavior towards the other sex. Presently, Marine Corps policy assumes that males are the perpetrators of sexual indiscretions, now *de facto*, crimes. My data, however, indicate at least equal culpability of the females involved.

Alcohol + Sex = Rape[31]

Leora Rosen of the National Institute of Justice specifically addressed rape rates and military personnel in the U.S. as an exploratory study at a state-level analysis. She tests Baron, Straus, and Jaffee's cultural spillover theory, which asserts that the more a society tends to legitimate the use of violence to attain ends for which there is widespread social approval, the greater the likelihood of illegitimate violence.[32] Her hypotheses were that the correlation between rape rates and military personnel would be higher in communities associated with Army and Marine Corps personnel because of their more direct association with combat *and* lower in those communities associated with Navy and Air Force personnel.[33] Ultimately, her study found no support for the cultural spillover hypothesis and that there were no significant correlations between rape rates and the presence of military personnel.[34] She found, however, a correlation between rape rates and alcohol consumption,[35] a correlation acknowledged by the U.S. military as well.[36]

Similarly, Venia Ceccato, et. al. embarked on research into a potential geographic relationship to rape incidents. The research affirms that rapes typically occur spatially within geographic spaces shared by perpetrator and victim, spaces with the characteristics of easy access (e.g., density of potential targets), rhythmic behavior patterns, and the absence of a capable guardian (e.g., isolated or restricted visual and auditory surveillance). Ceccato also found that leisure activities that include alcohol consumption directly correlate to forcible rape.[37] These data support the findings of Rosen and Ceccato, et. al. involving alcohol consumption. Although qualitative, these data corroborate these findings, adding nuance through description. Sgt Reese, for example, describes an incident involving alcohol and alleged rape:

> We had a few rapes that I know of. One was with a girl who was underage, drinking, already on restriction, but she had a male visitor. Supposedly, the male was there to pick up her roommate for a date, but they started drinking together and had sex. The roommate walked in on it. Later, the girl claimed she was raped. I don't know what really happened in that BEQ room…no one did and there were a lot of confused and angry people after that happened. The guys hated all of us females, and we hated her for putting herself out there. Now, the way it is, is that if the female has *any* alcohol in her system and she makes an allegation, it's *rape*. All I know is that there's no negotiating that fact. Per Marine Corps Order, alcohol plus sex equals rape. Period. I don't know how that can really be, like in the real world with married Marines and all but, per the order, it's a fact. It just happens to be a fact that makes absolutely no real sense.

Major Margo describes her own bad judgment while a young female Marine in Okinawa mixing sex with alcohol:

One of the stupidest things of all the many stupid things I did on active duty was as a Lance Corporal in Okinawa. A couple of WM friends and I went to an outdoor concert at Torii Beach where we met a couple of male grunts from [Camp] Hansen. We drank saki and beer and God only knows what else. My friends decided to go back to the barracks around dusk, but I hadn't had enough partying yet, and the guys said they'd take me back to base later. They didn't. I woke up the next day, about mid-morning on a Monday, in a trashed Futenma hotel room, physically brutalized. I am sure we, at least the three of us, had had sex utilizing every orifice of my body numerous times and, of course, I was horribly hung over, too. I had no money, no clothes, no ID, and absolutely no clue what had happened in the previous 18 hours. I had missed morning formation, PT, and was now naked and UA [Unauthorized Absence] from work. I had no one to blame but myself for putting myself in such a dangerous situation. I knew better! There had been an incident a few years earlier where a couple of Marines raped, murdered, and burned the body of a female Marine companion under similar circumstances. The pressing question at the time, though, was how to get myself out of the trouble I was already in with my unit. I don't even remember what I did. I am sure I tried to weasel out of trouble by going to the BAS [Battalion Aid Station], but I am sure the bruises and alcohol breath didn't fool anyone. I thank God they didn't rough up my face. That afternoon, I went up for office hours. I kept my crossed rifles, barely, but I got a nasty little entry in my SRB [Service Record Book], and they stuck me immediately on mess duty for all of November, December, and January. Of course, I was worried about STDs and, even though I was on the pill, pregnancy. Fortunately [Laughs], the guys I was with were way out of my normal circle of acquaintances so, as far as I know, no one even found out the truth of what happened. I guess the irony there is that even *I* will never know what happened. I don't blame them for what happened. Who knows? Maybe in a drunken passion I consented and asked for what I got. I did have the sense to be grateful to be alive. The pain and bruises eventually went away, but I can't erase the knowledge that I did something so dangerous and so stupid. I learned, though, and have never allowed myself to get out of control like that again.

GySgt Claire describes a situation that ruined a nice little perk in the desert that everyone enjoyed until one individual ruined it for everyone:

> We'd have fifty-cent draft night at The [Twenty-nine] Palms. Some guy went there one night and got really wasted, and he went around to different rooms, trying to find some sex. One room he tried, the girl bitched him out and told him to get lost. Another door he tried, apparently, the girl was asleep with the hatch unsecured and he raped her. He blamed the rape on fifty-cent draft night, so that ended immediately. It was just a criminal blaming something or someone else instead of owning up to making stupid choices. The guy had a history of alcohol abuse. He had a history of violence against women. But, it was the cheap beer that one night that was the "problem." Therefore, the hundreds of responsible Marines who enjoyed fifty-cent draft night every week and didn't commit a criminal act were punished because of one asshole who did.

Captain Taylor describes the individual nature of sexual behavior in the Marine Corps exemplified by a situation in which she found herself in the mid-90s:

> I did have one sexual harassment issue in the 20 years that I've been in the Corps. I was a young Marine. I was working a booth at some kind of event andwas groped by a Gunnery Sergeant who was drunk. I was pretty pissed off because being drunk is no excuse for bad behavior. It wasn't talked about, though, until another female Marine mentioned to me that he had done the same thing to her that day. We could have ruined his career but, by that time, he was retiring anyway. It just didn't seem worth the effort to prosecute him. We knew that doing anything legally wasn't going to change his mind about what he'd done. We were offered the opportunity to have counseling. Can you believe that? What a *joke*! Hell, instead of counseling, I wanted to show up at his door and tell his wife what he'd done and let her deal with him! If anything, *that* might have given him the kick in the balls he needed!

Supporting RAINN's theory, these stories and many others of female Marines attest that *individuals* behaving badly, not a failure of leadership or the presence of a "rape culture," comprise the criminal manifestation of sexual harassment, assault, and rape in the Marine Corps. Captain Taylor further states the conclusion most succinctly, "This stupid incident didn't affect my opinion of the Corps of Marines, in general. This was one moron acting stupid because he was drunk. It could have happened just as easily at a church bazaar or a school fundraiser. Men just do stupid things." According to all female Marines interviewed, leadership principles guide Marines and provide positive motivations; however, they cannot prevent individuals, male or female, from making bad decisions. Even as U.S. Marines, people are allowed the freedom to act on their impulses, for better or worse, both male *and* female.

Relinquishing Privileges of Rank for "Protection"

Numerous female Marines interviewed described being in situations where a patronizing male SNCO, officer, or the USMC institution itself expressed that females are not competant to ensure their own personal safety against sexual assault. Although the Marine Corps Martial Arts program professes self-security training the data assert that, regarding "One Mind, Any Weapon," all bets are off if that "One Mind" is connected to a body that does not include a penis regardless of her belt color. Further, MCO 1752.5 of 28 September 2004, "Sexual Assault Prevention and Response Program" condescendingly states: "Leaders must be aware that Marines who are sexual assault victims are physically, mentally, and emotionally traumatized and wounded. *A wounded Marine must never be left on the battlefield.*" [Emphasis added.] It goes on to say, "Marines are our most precious assets."[38] The very language used reflects postemotionalism, the Marine Corps now cast as a therapeutic society, Disneyesque, and artificial.[39]

Regardless of the Order's questionable language, no disrespect is intended here nor minimization of the experience of rape victims such as the two described by one female officer below. These examples merely serve to illustrate contemporary Marines' confusion over appropriate male-female behaviors as well as the contradiction of the Marine Corps' attempts to support females as "Marine leaders" while also casting them as vulnerable and potential victims of sexual crime. Specifically, Captain Sharlene describes relevant situations in both Iraq and at CAX in Twenty-nine Palms:

[In Iraq] there was one female who was raped. She was dragged into the head at gunpoint by another Marine. I talked to my Marines about it and one of the female Marines said, "How do they know it was rape? Some people like rough sex." I was appalled by this and the attitude among the other Marines, male and female, was similar. They blamed the victim and asked questions like, "What was she doing there? Why didn't she fight back? What was she wearing?" It was so disheartening. After that, the command's response was that females in the camp were not allowed to walk alone after dark. Women had to either be in twos or escorted by a male. They wanted *me* to do this, a Marine officer with no female peers on the base! I was supposed to call a male corporal to "escort" me to the head at midnight if I had to go...? There was no fucking way!

[At CAX] a Marine followed a female into the shower and raped her, so the response was to issue whistles to all females. There's no way in hell I was going to advertise my supposed vulnerability around my neck. And, that became a real joke, too. My peers would even say, "Hey, Barbee, where's your whistle? Hardy-har-har. You better get your whistle." It was so demeaning I couldn't believe it. My response was, "Fuck you, idiot." In sum, instead of trying to get to the heart of the matter, which was identifying and prosecuting the criminal behavior of certain males, the response was to emphasize the vulnerability of all female Marines. It was so wrong on so many levels, I still cannot believe it. And, of course, the effort was not put into us being able to protect ourselves, but the emphasis was put on the fact that we should call the males to protect us! Another female lieutenant, who was a green belt, and I actually got together with a bunch of the enlisted females and talked about what we could do to combat predatory males. Go for the nuts. Go for the eyes. Basically, just tear them a new asshole because that's what's required. But, we were criticized for that, too, by the male leadership. "That's not your job." "Who do you think you are?" So, we found out that, when it comes to gender issues, we were females, *first*, and Marines second. Oh, and forget about being an officer because little ol' me might need a male lance corporal to take care of me. And, I was *ordered* by a male 1stSgt to attend a meeting for females about how to protect ourselves. I went as a show of leadership for the other females, but resented it then and to this day. He was enlisted, but he was male, so *he* had the right to tell *me* what I needed to do for my own safety and for the safety of my female Marines! It still makes me so goddamn mad that I'll never forget it.

From an evolutionary perspective and in contemporary society, it seems that a female's sex is commodified initially by nature as a reproductive asset and then by society as an item of value vulnerable to males' interest and desire identified previously by Buss and Blau. Evolutionarily and socially, it is an asset to be protected from males *by* males. In our postemotional society, however, the social and sexual freedoms females have achieved place them in a precarious situation of how to actually manage their sexual and social freedom against the evolutionary backdrop of being favored as biologically reproductive beings, sought after by males, sometimes aggressively, and not always by the males they are willing and eager to mate with. Women's sexuality, it seems, renders them vulnerable to predation, which must be mitigated.

Willful Virgins: Female Marines Choosing with Whom they Mate

As portrayed in previous chapters, female Marines often revel in the opportunities for sexual opportunity in the Marine Corps while fully understanding the potential impact of their choices on their social reputation and military career. They also express a keen awareness of many of the behaviors enacted by those males seeking to mate with them. In the American civilian community, females are pursed by male suitors in a mating environment where the sex ratio is roughly 50-50. Female Marines, however, with the skewed 7% female to 93% male environment can find the process overwhelming. This research asserts that successful female Marines, confident in their competency socially and professionally, develop ways to discourage overbearing and unwanted males. For example, GySgt Claire describes thwarting one predatory males' sense of ownership:

> I was on a date one time and another guy came up. I don't even know if he was a Marine or not, but he came up to me and wanted to dance. I said, "No, thank you. I'm here with him." He said, "Well, he's a white boy, and I'm gonna kick his ass because he ain't supposed to be with no sistas." I said, "Okay, you kicking his butt doesn't mean you're gonna be with *me* so just take your

happy ass on down the road." It really pissed me off. He thought he was just gonna remove the competition and I would be his. Uhhh....no. It doesn't work that way anymore.

Similarly, Captain Kami describes an overbearing boyfriend:

If anything, I think being a Marine makes me less attractive to a man because they see me as their equal and I think that's *not* what so many of them want in a mate. I think they don't want a strong, independent woman as a wife. I think they still want to call the shots and the little missus to say, "Yes, sir." In a recent relationship, I'd been dating a former male Marine for about five months and I was on deployment and wanted to stay on deployment, which I did, and he got all huffy and said, "You didn't discuss this with me!" It was kind of funny because I had no ties to this guy. I was in Northeastern Africa, which was the most incredible duty, and he wanted me to come home and play house! No fucking way! And when he said the "magic" words, "If you really loved me, you'd be with me," I just totally shut him down. Un-uh. Nope. That was that. If you're that needy, go home to your momma.

Numerous similar stories of female sexual independence and empowerment could be included here; however, these two succinctly represent the many female Marines who realize that, like Frey's Willful Virgin, their key to "freedom" is retention and selective dispensation of their sexuality. It is their key to independence. Even in sexual relationships with males with whom they choose to mate, they continue to view their sexuality as their *own*, not the possession of a male. Many recognize that relinquishing it potentially transfers its power in the relationship to the recipient as Blau espouses.

Freedom & Power within the System

If it has not become evident to this point, it must be stated clearly that, far from being victimized by a "misogynist" and repressive institution, female Marines display a wide range of social, sexual, and occupational freedoms, many more than previous generations. However, such freedoms come with a stipulation because females remain an oddity in the Marine Corps, an alluring 7% of sought-after sexuality. They have the opportunity to use, suppress, or misuse their sexuality in a mating environment heavily skewed in their favor. As described by the female Marines who live in this environment every day in a wide variety of leadership positions and occupational roles, female Marines are certainly cognizant of their surroundings. As asserted by Christine Williams, they are fully aware of the males' sexual attention. Some are unsure of how to appropriately deal with this attention, but *all* must choose to respond in either appropriate or destructive ways. As a Marine, this is an occupational imperative.

Unlike previous generations such as the Free a Man to Fight female Marines who lived in a separate sphere from the males where appropriate conduct between men and women was more strictly codified in American society and the Marine Corps, AVF females dealt with a nameless sexual harassment which resulted from females storming the males' bulwarks, entering previously closed MOSs and invading their barracks, shops, and heads [bathrooms]. The adaptations those females developed were varied, but most assert that their experience in the Marine Corps was positive once they established mutually respectful relationships with their male counterparts. In the present-day Marine Corps, however, postemotional attitudes cultivate an environment of confusion, indiscretion, and resentment related to behavior between the sexes. In this perplexing milieu, men and women, as individuals, continue to make naïve or manipulative choices that reflect negatively on the organization and everyone within it.

Patriarchal Benevolence, Females' Sexual Security & Relative Power

Over four decades, radical feminists demanded sexual freedom for females believing female ownership of sexuality to be the key to social freedom and equality with males. They have demanded the same sexual opportunities as men, to pick and choose sexual partners and engage in recreational sex without criticism. In doing so, they cast off the restrictive mantle of male and familial (i.e., kinship) protection, whether in the form of a father figure, a brother, or another appropriate chaperone. This creates a pseudo-level playing field where females are equally "free" to be sexual, but the consequences of that freedom remain compromised. In reality, although a female positions herself as a physical, sexual equal to males, she remains physically weaker than most males. Historically, to correct that physical imbalance, society has determined, demanded, that males behave in an "appropriate" manner via chivalry, ethics, or laws to protect female sexuality from encroachment by unapproved males. What radical feminists sought to accomplish under the Constitutional banner of "equality" merely exchanged one form of patriarchal benevolence for another: in a Weberian twist, the charismatic and social kinship system of female sexual protection by male kin has merely been replaced by the authoritarian system of protection by law.[40]

In Meštrović's postemotional context, where individuals choose to claim victim status at will, an innocently intended romantic advance can easily be perceived as threatening or, worse, used as leverage against the potential suitor and the establishment. It is true that sexual predators do exist in American society, which remains a serious concern. However, the very subjectivity and ambiguity of "sexual harassment" as a social and legal relationship management strategy in the military lends itself to be used as a tool by the disenfranchised or manipulated by the marginal to feel powerful.[41] In mating, a female Marine has the power to acquiesce, negotiate, or allegate. The power is *hers* to be used or misused per her whim. The United States Marine Corps as an organization and the actors within it throughout the chain of command must more rigorously identify, investigate, and mitigate this imbalance and potential misuse of female sexual power.

Endnotes

1   Mady Wechsler Segal, "The Military and the Family as Greedy Institutions," *Armed Forces & Society* DOI: 10.1177/0095327X8601300101 (Fall 1986): 12.

2   Brownson, "Battle for Equivalency," 3-4.

3   Nix, "American Civil-Military Relations," 93.

4   Mitchell, *Women in the Military*, 194.

5   Brownson, "Rejecting Patriarchy," 3.

6   Buss, *The Evolution of Desire*, 15.

7   Ibid, 5.

8   Brownson, "Battle for Equivalency," 6.

9 Blau, *Exchange and Power*.

10 Buss, *Evolution of Desire*, 211-214.

11 The chain of command and the UCMJ exist in tandem and for numerous reasons, one of which is a checks-and-balances system against abuse of individuals within the organization. The statement made here in no way implies that there are no "power plays" among Marines, as almost all exchange relationships occur amongst and between persons "unequal" on at least some level. It does assert, however, that coercion of a criminal nature occurs in extremely rare, one might say "intimate" circumstances, such as sexual assault and rape, which are criminal offenses and, as such, fall under the jurisdiction of the UCMJ.

12 Wendy Chapkis, *Live Sex Acts: Women Performing Erotic Labor* (New York: Routledge, 1997), 22.

13 Winerip, "Revisiting the Tailhook Scandal."

14 Madeline Morris, "In War and Peace: Incidence and Implications of Rape by Military Personnel," in *Beyond Zero Tolerance: Discrimination in Military Culture* eds. Mary Fainsod Katzenstein and Judith Reppy (Oxford, England, 1999): 173.

15 Brownson, "Battle for Equivalency," 5-6.

16 Buss, *The Evolution of Desire,* 9.

17 Michael V. Studd, "Sexual Harassment," in *Sex, Power, Conflict: Evolutionary and Feminist Perspectives*, ed. David M. Buss (New York: Oxford University Press, 1996), 84-85.

18 Buss, *Evolution of Desire*, 8.

19 Johannes Han-Yin Chang, "Mead's Theory of Emergence": 414.

20 Brownson, "Rejecting Patriarchy," 3.

21 Meštrović, *Postemotional Society*, 15.

22 Zimmer, "Tokenism and Women in the Workplace," 71-73.

23 Kümmel, "When Boy Meets Girl," 619-620.

24 Meštrović, *Postemotional Society*, 16.

25 Buss, *Murderer Next Door*, 116.

26 Carin Perilloux, Judith A. Easton, and David M. Buss, "The Misperception of Sexual Interest," *Psychological Science* doi: 10.1177/0956797611424162 (18 January 2012): 146-150.

27 David Buss, *The Murderer Next Door*, 63-64.

28 Ibid., 88.

29 Valli Kalei Kanuha, Patricia Erwin, and Ellen Pence, "Strange Bedfellows: Feminist Advocates and U.S. Marines Working to End Violence, *Affilia* doi: 10.1177/0886109904269053 (Winter 2004): 358-375.

30 "RAINN Urges White House Task Force to Overhaul Colleges' Treatment of Rape," Rape, Abuse &

Incest National Network (RAINN), NewsRoom, https://rainn.org/news-room/rainn-urges-white-house-task-force-to-overhaul-colleges-treatment-of-rape.

31 For preliminary research on the association between alcohol consumption and sexual assault, see Antonia Abbey, et. al., "Alcohol, Misperception, and Sexual Assault: How and Why are They Linked?" in *Sex, Power, Conflict: Evolutionary and Feminist Perspectives*, ed. David M. Buss (New York: Oxford University Press, 1996), 138-161.

32 Larry Baron, Murray A. Straus and David Jaffee, "Legitimate Violence, Violent Attitudes, and Rape: A Test of the Cultural Spillover Theory," *Annals of the New York Academy of Sciences*, doi: 10.1111/j.1749-6632.1988.tb50853.x 1988, 528: 79–110.

33 Rosen, 947.

34 Ibid., 952.

35 Ibid., 956-957.

36 United States Commission on Civil Rights, "2013 Statutory Enforcement Report: Sexual Assault in the Military," September 2013, Washington, DC.

37 Vania Ceccato, Robert Haining and Paola Signoretta. "The Nature of Rape Places" *Journal of Environmental Psychology* 40 (2014): 97-107.

38 Marine Corps Order 1752.5, *Sexual Assault Prevention and Response Program,* 28 Sept 2004.

39 Meštrović, *Postemotional Society*, 91.

40 Weber, "Legitimate Domination."

41 Meštrović, *Postemotional Society*, 16.

Chapter Eight

Personal Lives of Female Marines: Appeasing Greedy Institutions

Military historian Martin van Creveld asserts, "Perhaps it was to the credit of American women that most of them refused to follow the call of their would-be leaders and believe that they could have their cake and eat it too. Faced with the choice between maintaining their *traditional privileges* and obtaining equality, they preferred the former causing the ERA to be abandoned."[1] [Emphasis added.] What he fails to acknowledge and discuss is that "traditional privileges" were nothing more than domestic prisons or, perhaps more appropriately, breeding agreements with the "privilege" of a lifetime sentence of servitude to home and family. Historically, promises of domestic bliss through sex/gender "equality" prove to be lies, even if well-intended.[2] Female Marines often find themselves trapped by their own liberation and occupational success. In additional to the "traditional privileges" expected of women to provide familial nurturing, foraging, food preparation, and cleaning functions, they must also continue to perform superlatively as United States Marines.

Buss asserts that fulfilling each other's evolved desires is the key to harmony between a man and woman. A woman's happiness increases when the man brings more economic resources to the union and shows kindness, affection, and commitment. A man's happiness increases when the woman is more physically attractive than he is, and when she shows kindness, affection, and commitment. Those who fulfill each other's desires have more fulfilling relationships.[3] Apparently, if both sexes want "kindness, affection, and commitment," the divisive attributes are a female's desire for the male's economic resources and a male's desire for the female's attractiveness. Evidenced in previous chapters, female Marines exhibit prowess and success economically, physically, and professionally. They are committed, successful Marines and still exhibit kindness and affection in military and social kinship relationships. In theory, they do, indeed, "have it all" from an evolutionary perspective. However, the Marine Corps' litany of military requirements and expectations in addition to just "doing one's job" every day while embracing and enacting the roles of wife, lover, mother, friend, and a myriad others, requires extraordinary self-sacrifice. The notion of self-sacrifice is not alien to women; women understand this concept all too well.[4] In this challenge as well, female Marines are enigmatic, caught between worlds as women and warriors.

The Marine Corps 24/7/365: Whiners Need Not Apply

As previously described, Marine Corps life begins in Boot Camp or OCS, expands through MCT or TBS, and then a female Marine attends her MOS school to learn her actual military occupation. Once in the Fleet Marine Force (FMF), the day-to-day expectations only increase. There is required unit PT, usually very early in the morning before working hours. In Okinawa, for example, the typical summer PT formation is at 5:30 a.m. with Marines on the road and ready to go well before sunrise and its accompanying crushing heat and humidity. Usually monthly, Marines stand duty, a 24-hour watch, at their barracks, workplace, or elsewhere on the base. Annually Marines re-qualify with the rifle or pistol or both and swim re-qualification is required depending on how proficient one is in the pool. Added to that are required Professional Military Education courses, affectionately known as "PME," a schedule of schools to attend according to one's rank, such as the Corporals Course, the Sergeants Course, the Staff NCO Academy, and the Advanced Course on the enlisted side. For officers, career courses, Expeditionary Warfare School (EWS), and Command and Staff College (CSC) are required, and the list goes on. In the interim, there are PFTs to run and proctor, Thursday field days to conduct or supervise, urinalyses to take or monitor, dog-and-pony shows, changes of command, Marine Corps Birthday Balls, trips to the field for training and, of course these days, deployments to combat zones. The Marine Corps is no 8-to-5 operation and any comparison to a civilian equivalent, including that of first responders, consistently fails.[5] Marines are not "first responders," they are "constant responders."

The Marine Corps exemplifies Coser's ultimate "greedy institution," exercising pressures on component individuals to weaken their ties, or not allow formation of any ties, with other institutions or persons that might make claims that conflict with its own demands.[6] From the data, it cannot be denied that service in the United States Marine Corps in war and even in peacetime demands participation by a different kind of person, especially a different kind of woman from Anytown, USA. In the words of 1stSgt Halona, "There really needs to be that point where some people need to be told, 'Sorry, but you're just not good enough. Thanks for playing. Goodbye.' I think the government spends a shitload of money on people that just shouldn't be in the Marine Corps." This discussion so far introduced numerous personas, scenarios, and reasons why females join and remain in the Marine Corps. The focus now turns to the day-to-day drudgery and spontaneity of the lives of female Marines.

Civilians' Limited Understanding of the Reality

Many authors writing "women in the military" books consistently prove that they have minimal, if any, understanding of how demanding the military lifestyle can truly be. They often choose to focus on the occupational aspects rather than the engulfing lifestyle indicative of the Marine Corps.[7] Female Marines, after "work," more often than not, do not return home to family and a cadre of established friends. She might be in Diego Garcia, Chad, Baghdad, Okinawa, or at sea with only her fellow Marines *as* her family for months at a time. To succeed, female Marines must embody intensity, commitment, physicality, and passion that most female Americans either do not possess or choose to sequester to better "fit in" with mainstream American society.[8]

In some respects, female Marines are quite "normal" women, while also being quite extraordinary in their attitudes, behaviors, and achievements. If dissatisfied with one's job, female Marines simply do not have the option to submit a resignation and go home. Recognizing their enigmatic status, if given the option, the data indicate that the successful female Marines don't want to. The mosaic of stories presented here includes numerous and varied references to "home" and the definitions of "home." Regardless of the geographical location of home, the roles and behaviors associated with home remain.

As presented previously through a variety of scenarios, male-female, civilian-military and, of course, relationships within the rank structure must be negotiated everywhere a female Marine *is* geographically and at the time of her personal life and career. For example, Captain Dawn describes her relationship with other incredible female Marines and the unique experience of females who choose to live outside the boundaries of what "normal" Americans, both male and female:

> All my life, I've had male friends, but now I have all female friends, all officers and SNCOs. They're the best friends I've had in my entire life. It's not just because I'm older and I've begun to value this type of relationship. It's really a different *kind* of relationship entirely. I trust these women with my life. If I had any, I'd trust them with the lives of my children. They care. They're *committed*. And, most of all, they're smart and they know how to get the mission, whatever it is, accomplished. I believe there's a difference between those people who have done amazing things and those who have not. I would count all of these women in the group of truly amazing. Getting up and going to work every morning is not "amazing." Getting up knowing that you have orders to Iraq and packing up your kids to go live with your mother and leaving them with a confident smile, not knowing if you'll ever see them again, is what takes guts. We, as female Marines, live in extraordinary circumstances almost every day. And, in saying this, I can't slight the women and men who have served exclusively in peacetime. Beirut, Panama, the USS Cole, and 9/11 are perfect examples of why choosing this life is unique. When you sign up, you *understand* that the world security situation can change in a split-second. We train to be ready for that and to expect it. If you can't deal with that, you really need to stay home.

Sgt Reese, a former undergraduate student at Emory University describes the uniqueness and determination of female Marines:

> I know I am and my female Marines friends are strong-willed. We're good natured for the most part, but don't piss us off. I think we're more committed and determined than most of the men out

there because we *have* to be better. We have to make the really hard choices. They can expect their mothers when they're young or their wives when they're older to manage their personal lives, like paying the bills, maintaining a house, and raising a family. As female Marines, though, whatever our MOS or rank, *we* have to manage all that *and* be 100% a Marine. There's no one to pick up the slack for us. *We* have to be it all!

In the wisdom of her youth, LCpl Patrice also warns the uninitiated of the costs demanded by the Marine Corps lifestyle:

It takes a certain kind of person to be a Marine. It takes a lot of flexibility. It takes someone who can swallow their own personal pride to achieve the greater pride of the group. It takes courage to speak up when you should and to shut up when you should. It takes wisdom to know when to do each of these things. It takes honesty and self-knowledge. You have to admit when you're wrong and recognize when those around you have done something outstanding and honor that. It takes strength of character. Marines are different from other people because of these things. If you're not going to uphold these standards, don't apply. Or, if you find yourself in it and can't do it, do your time and get the hell out. It takes 110% and if you're not up to it, please don't apply 'cause we don't need you.

Sexual Double Standards

Separate from a woman's need to perform professionally and competently in the Marine Corps, she also faces the expectation of appropriate conduct as a woman. As discussed previously, sex-based double standards remain when it comes to behavior out of uniform as well as in uniform.[9] Captain Caroline speaks blatantly to the issue of sexual double standards:

Do we hear that Joe or Frank or Charlie is a "slut" and, therefore, unworthy of being a Marine officer, unworthy of being a leader? Hell, no, we don't! This sexual double standard is probably the worst element of the Marine Corps that I've witnessed because, when we frame sexuality and leadership as synonymous elements, all females are screwed, no pun intended, and that's the way it's been with men against women for centuries. Can we change it? I don't know, but that is the element that must be changed for females to be successful in the Marine Corps. There are constantly scandals, issues, and allegations in the Marine Corps. There are lieutenants sleeping with enlisted, and instructors sleeping with students, SNCOs sleeping with officers' wives. It's the best soap ever written but never screened. A lot of this is because of the Puritanical attitude of the Marine Corps towards sex. It's funny because male virility is so paramount, but the actual sex is considered "dirty" and "inappropriate" unless it's in the perfect context which I imagine is "the lovely young Marine officer marrying the virginal young female co-ed debutante." It's fucking crazy! How often does *that* happen?!? These officers' wives might have been the town pumps, but you put them in this new environment, and their supposed "beauty and grace" take on this whole other fairy tale meaning. It makes me retch! These women are no different from me except that I wanted to *be* a Marine and not just live my life as a baby-making hanger-on like them.

Although Brian Mitchell asserts that, "Chivalry honored women with care and safety if not freedom, at least the freedom to be men." Today, women are free to live as coarsely and brutally as men, while men are 'desensitized' to the suffering of women in training. Yet, somehow, when women discern the slightest offense, the old ways are always to blame."[10] Although he does not identify the "old ways," it seems apparent that, similar to van Creveld, Mitchell is referencing the male ownership of females' sexuality that earned females the "honor" of protective chivalry, benevolent patriarchy. Many female Marines politely reject "chivalry" in the work place if not in their home lives, often striking a path of power in one realm and submission in another as she chooses. They choose equivalency and the give-and-take necessary to maintain satisfying relationships in both realms.[11]

The Feminine Warrior

Although seemingly an oxymoron, similar to Christine William's finding in the 1980s, contemporary female Marines also largely, "…associate femininity with dignity, bearing, and self-confidence."[12] The key, however, is to balance the roles one adopts in accordance with the requirements of the situation in which one exists, either as a lady or as a Marine. For example, SSgt Tanisha describes her situation:

The Marine Corps never took anything from me, especially not my femininity. I get my toes done. I get my nails done. I keep myself manicured. I wear make-up. Sometimes, I get my hair done. I never let myself look too beat down or too masculine. I definitely maintain my femininity. At first, I didn't know how to do that and I talked to one of my Drill Instructors and she said, "Look. The Marine Corps is gonna be the Marine Corps. You be yourself *in* the Marine Corps. Don't try to be like anybody else. If you're a girly girl, be a girly girl. Just do what you need to do." 'Cause I did struggle with that at first. I thought they were gonna take *me* away from *me*. So, when she told me that, I knew that I could do that. So, I put the uniform on, and do what I gotta do. You learn early on that you don't embarrass Marine Corps or embarrass yourself. You and the Corps become a package deal, each reflecting on the other.

Similarly, SSgt Robin describes the seamlessness of her experience as a female Marine:

Too many women, especially girls and young women, follow whatever the media dictates that they should be. It's sickening. You want to be *thin*, but not be physically and emotionally fit and strong? It doesn't make sense. You want to be the CEO of a company, but you have to wear a tight, short skirt, or even worse a masculine business suit, to make that happen? Why?!? I think a lot of stellar female Marines are very humble. They don't try to "act" outside of what they know, what they believe, and what they're comfortable with. Female Marines aren't frauds because there are too many people just waiting to call you out. You better be able to put up or shut up. I have my make-up compact and I have my cammie make-up kit. It's no compromise for me to wear one or the other. I am totally comfortable in either. It's either time to get slinky, or it's time to go to war. I'm never compromised because I am both of those people: woman and warrior. I never have to be a dude. My Marines still call me "Staff Sergeant" when they see me in a sun dress and sandals. My roles are seamless and no part of me suffers because of the other.

GySgt Tiffany, too, revels in the duality of being a warrior princess:

Female Marines need as much support as anyone else. Sometimes, we just want some polish to do our nails, and then not be criticized for doing them. Everyone needs to support the troops beyond any political agenda. My friends have died out there for your right to be unhappy. If I die, I don't want anyone to mourn what I have done. I have too many friends who now reside at Arlington [National Cemetery]. I love being a Marine. I love that respect that I get from other people and the surprise that I get from people. "What? You're a Marine?" I love being able to wear cammies all day and go home at night and put on a silk dress, and there's not a person in my daytime world that would recognize me. I love the duality of walking around in a man's world and beating them up, and then I can go home and be a princess and be taken care of my man. I don't have to *always* be hard. I don't have to be a mongrel of a woman. I can be a Marine, and do my job and be good at it, and I can still be a lady. I think a lot of people forget that we are ladies, too.

Captain Gail describes her experience of skewed gender roles in her marriage to a male Marine officer:

There are times when my husband seems to take on a more feminine role than I do. Like, he'll get very upset about something, and I tend to take a more masculine attitude. He'll continue to harp and I'll say, "We talked about this. *Why* do we need to keep talking about this?" It's really funny. I think, in some ways, I do have what might be considered masculine thought processes and attitudes, and that's helped me be successful in the Marine Corps because it is a male-dominated organization. Coming to 4th Battalion at Parris Island, I remember thinking, "Wow. Do I want to put myself in an environment with all women?" But, it hasn't been a problem at all. The women I work with are professionals and are some of the greatest Marines, in general. I love to see them getting the job done and getting promoted and being placed in the leadership roles that they're in. These women are not butch by any means. They're attractive, competent, and feminine women. However, when it's time to get the job done, there's no dicking around. They get the job done. They're the perfect combination of what is valued in both sexes without compromising who they truly are. They're wives and mothers. They're riflemen and engineers. Being one "thing" doesn't preclude you from being any other "thing," especially in the Marine Corps where so much is demanded of people, physically and emotionally. I've seen male Marines pack a duty hut to watch television. Was it a football game they were watching? Hell, no. It was a damn soap! And, they're getting all emotional and misty-eyed over it! [Laughs] Did that make them feminine or less of a Marine? No, not at all. I laugh, yeah, but only to myself. I love these silly guys. Just like a female can be a Marine and still be very feminine and attractive, they can be emotional, too. It's only fair. Neither sex has the exclusivity to be one thing to the exclusion of the other.

The Marine Corps Mating Game

Perhaps not surprising, many female Marines express unique challenges in dating and maintaining romantic relationships. Some attribute the difficulties to the stereotypical labels previously identified. Others attribute relationship challenges to the Marine Corps lifestyle itself, such as the high operational tempo, deployments, and frequent relocations across the country and around the world. For many female Marines, the dating game is much more complex than for civilians because of their status as a Marine and the ensuing occupational and social empowerment and confidence they embody as identified above related to Buss' theory.

Numerous female officers describe their frustration at "playing the mating game" in the Marine Corps. However, not a single enlisted female described being in a similar situation. With females comprising 7% of the total force and surrounded by an abundance of potential male mates, it seems that mating would be an extremely easy endeavor with females able to choose from a range of available mates. Fraternization, however, forbidden by the UCMJ, targets the egregiousness of sexual indiscretion between the ranks, compounding and confirming the tenets of evolutionary psychology. In essence, sexual liaisons between actors enacting disparate levels of power within a hierarchy disrupt good order and discipline.

The limited pool of acceptable mates challenges female Marine officers. Women worldwide prefer to marry up, which stymies female officers' mating success.[13] For example, Iraq combat veteran Major Amy describes the difficulty she experiences with the mating game as a female Marine officer:

[Laughing] People with a single address for more than six months have personal lives! Honestly, though, I want to go on MSG duty, so that's out of the country for two years. I've spent seven years in the Fleet and been on five deployments in that time. If I could be at MCRD San Diego as a female Marine officer, like a series commander, I would do it in a heartbeat. At Parris Island, though, I would be a single female lieutenant colonel, and that is just sad and frowned upon on so many levels. The attitude is, "Sure, you may be successful in the Marine Corps, but you've failed at the mating game," which is really a woman's goal/role anyway, right? I have dated civilian guys, but I want a guy that's strong and he'll have to understand that I am a Marine and that I'm not going to change to play house. If that's what I wanted, I could have stayed home after high school. It's funny, too, because when I meet the wives of my fellow male officers, the economy in the relationship is often so skewed. I want a relationship that's more balanced instead of having, like the males, a spouse that caters to my wants and needs, follows me around, whines and cries when I'm gone, and takes care of my children, personal and business affairs because it's not *my* job cause I'm too busy doing "Marine" things. I could never be a stay-at-home mom. Maybe that makes me a "bad" female. I couldn't do the officer's wives club stuff because I find all that really vicarious and boring. So, where does that leave me? It leaves me a seven-year single female Marine major, hoping to have some down time and find a wild and crazy guy who'll put up with me for the rest of our lives. [Laughs] He's out there. I know he is.

Similarly, Captain Kris, a well-established fiscal, procurement, and contracting officer with a MBA, assesses her status as a single female Marine officer:

Dating as a 34-year-old female Marine officer is *not* easy! To be 34 as a single, female Marine officer is like being the old washed-up spinster in the civilian world. You're *career*, read "life," is more than half over and you're just *old* at 34! That's how it's perceived. It's crazy, but that's the reality. I went to MOS school at Camp Johnson after TBS for seven weeks. They offered me a position after my tour here at Parris Island to go back up to Lejeune to teach and be the OIC of the Contracting School. I turned it down. I *don't* want to go to Lejeune. I'm a single female and, even here [at Parris Island], is miserable for me. I have no single friends. I have acquaintances here, but guys don't want to hang out with me.

I've never dated a civilian guy because I think there would be too much weirdness. I really think that this lifestyle is too foreign to anyone who hasn't already experienced it to really understand and accept the things we do and why we do them. Before I meet people who have heard that I am a Marine thay are often shocked when they meet me in person. They say things like, "Wow! You look like a girl," or "You're not at all what I expected." When that happens, I always want to ask exactly *what* it was that they were expecting, but I was brought up better, so I don't ask. I guess they have this perception that I wear a crew cut and have an Emblem tattooed across my forearm or my chest. I really don't know, but it's something I, and I'm sure other female Marines, deal with in day-to-day life. People *know* what a "United States Marine" is supposed to look like. When you add "female" to the picture, they just don't know what to do with that.

Another problem I deal with is that my male peers are married and most have small children. I will be on the major board this fall and there's a 90% selection rate, so I'll be really bummed if I don't get picked up. But, that creates another problem because it will be really, really awkward being a single female Marine major. It's almost unheard of among males, but to be an unmarried female is downright scandalous!

Recently, I was on a TAD trip, and I was in a conversation with a couple of other female captains. We didn't know each other or any-thing, but we were all TAD so we talked during breaks throughout the day. Well, both of them had just had a baby within the past year so they had a lot more in common than I had with either of them, so after the class was done, I went to the mall. Well, they came walking up and greeted me and said that they were shopping to buy things to take home to their husbands and babies. I said that I was just hanging out and shopping for myself because I didn't have a husband or a baby, and they acted all sympathetic and were like, "Oh, don't worry. It'll happen." *It was surreal.* One, it was not my goal to "have" a husband or a baby, and, two, I didn't need their sympathy and encouragement because my lifestyle wasn't in line with their expectations. To this day, I am amazed by the entire encounter. Apparently, they think I'm "unfulfilled." Unreal! But, even when I joined the Marine Corps, the first thing my mom said was, "Oh, no! Everyone's going to think you're *gay*!' She was really mortified by this. Of course, now my family and friends see me as having a healthy and happy, if unattached, heterosexual lifestyle and a successful career that I truly love. I've traveled and have lived all over the place, enough to learn languages and customs. I've got my MBA and I'm really happy with the things I've done.

Many female officers believe rightly that they are a threat to male Marine officers' masculinity and status within the organization, beginning at TBS and throughout their careers.[14] From an evolutionary psychology perspective, this makes complete sense. Regarding men's status and women's desirability ["beauty" in Buss' actual language, which is all part of the female's "package"], he states:

> Everyday folklore tells us that our mate is a reflection of ourselves. Men are particularly concerned about status, reputation, and hierarchies because elevated rank has always been an important means of acquiring the resources that make men attractive to women [and allow them to dominate other men]. It is reasonable, therefore, to expect that a man will be concerned about the effect that his mate has on his social status – an effect that has consequences for gaining additional resources and mating opportunities.[15]

In support of Buss' theories, female Marines, particularly officers, perhaps are simply too successful to be attractive mates within the military hierarchy and the civilian community. Interviewees articulate these concepts in many of their stories.

The Married Female Marine Officer

A primary concern of many female Marine officers is how to balance their successful status and career in relation to that of their husband or boyfriend so as not to create discomfort for him in his professional career. Balancing a marriage as two Marine officers is no easy task, and many female Marines find themselves enmeshed in roles that soundly contradict each other, wife and Marine officer, and the expectations that emerge from those roles. Often, they find themselves in a quandary, unable to discern which role is appropriate to don at which time. Captain Joy describes the challenges of being a female Marine officer in an overwhelmingly male environment as well as the difficult transition to be an officer's wife while still an active duty Marine officer herself:

> The hardest problem, I think, with being a female and being an officer is the spouses of the male Marine officers. There is no worse feeling in the world that I have ever felt than going to a social function and wondering, "Where do I belong?" I remember there

was a specific incident that really illustrates the weirdness that comes with being a female Marine officer. There was a party at my battalion commander's house, and I was the only female Marine, one of only three officers at the party, and *the* only female Marine officer. It was all SNCOs and officers, so I was about 10 years junior to all of the other officers there because they were all prior enlisted. So, I walk in and I'm, I suppose, just the picture of a single, petite, attractive 24-year-old in blue jeans and tennis shoes. One of *my* Staff Sergeants walks up to me and says, "Hey, ma'am, how are you? Can I get you a drink?" I say, "Sure. Thanks." Then, he says, "No problem. The hens are all over there." *What the fuck did I just hear??* I was *so* insanely pissed and at the same time, though, I was devastated, too. Until I talked to him about it that next Monday morning, I am positive that he never even realized what he had said to me. What was even more difficult after that was that, still at the party, when I walked over to hang out with *my* Marines that I work with every day, the conversation stopped and an awkward silence took its place. And, of course, there *were* the "hens," all standing there looking at me like, "Why are you talking to my husband?" Then, I tried to join their group, but I'm thinking, "I don't know what you're talking about with the key volunteers and the bake sale and the care package coordination…" I don't know anything about that stuff and I don't want to and I don't have *anything* in common with you! It's the worst feeling walking into that situation and just not belonging anywhere. That was when I was unmarried. Those wives *really* hated me.

It's actually better now, though, that I'm married, but it's still weird. I'll go with my husband to one of his events and all of his Marines know that I'm a captain, but he's a first lieutenant, a grunt, hanging out with his Marines and he'll be like, "Hey, all the wives are over there." So, even he's relegating me to my appropriate "place." I've tried to tell him that I have nothing in common with them! I absolutely hate it! One good thing, though, is that he gets shit around the base because I outrank him. All bets are off, though, in the marriage when he's the "man" in charge. Honestly, though, when it comes to the male-female thing at those gatherings, for lack of a better word, my attitude is that I can have a conversation with your husband more easily than I can have a conversation with you, another female. Especially at our age, the mid-20s through early 30s, the women don't want to see a pretty, strong young female interacting with her husband or boyfriend. So, I prefer to be totally anti-social, which sucks, but I've found that it's the easiest way to deal with all the drama. If they think I'm a bitch, I really don't care.

Captain Taylor loves that her husband, also a Marine officer, dotes on her being a Marine, but worries about the implications for respect within his command and the implications for his career, which harkens back to Buss' theory described in Chapter 4 regarding women shunning men who fail to command the respect of the group:

> It's so funny when I am out with my husband because he's always so eager to tell everyone that I am a Marine. I feel like I'm taking something away from him because he's being so proud of me. It's crazy, because he's such a stellar Marine; I don't know why he would want to boast on me so much when I feel rather un-exceptional. I guess I do have a reputation as being a hard-ass, but I like it. I just would never, ever want anyone to say he's "whipped" because of the way he talks about me. I would never want my status as an officer to detract from his. He doesn't seem to care, but I sure do.

These and other stories of conflicting roles permeate the married female officer experience and are minimally mentioned by enlisted female Marines. When an issue, many adopt a trial-and-error approach, tending to err on the side of caution in potentially awkward social situations. Others, however, make a conscious choice not to participate at all in functions sponsored by their husband's unit. This alternative, however, may skew public opinion *against* her because an officer's wife, in particular, has implied social roles to perform.[16] Although Moskos recognizes the role of officers' spouses in the military community and states that in the postmodern era those roles have diminished,[17] spouses who make faux pas create embarrassment for their Marine spouse as well as their commands. As with so many other aspects of her life and career, a female Marine must strike that delicate balance to appease greedy institutional expectations of her proper role as wife and Marine.

Male Civilian & Prior Service Spouses

Evidenced in the data, although much less common than Marine-Marine marriages, some female Marines marry civilians. Of these marriages, many constitute a female Marine married to a prior service male, primarily because they understand the level of commitment required by the organization. These men lived the experience, can understand it, and also are willing to adapt their lifestyle in deference to the demands of hers. Cpl Sonia, married to a civilian who is not prior service, describes her marriage:

> My husband is a civilian and he doesn't mind my career and, yes, I have decided that this will be my career. He's a motorcycle mechanic, so he can travel anywhere and be employable. He's a big, burly guy but he's not at all concerned about me being a Marine. He even says that he's especially proud of me because he'd never have the guts to do half the stuff I do. He's a great guy, and we're great for each other. When the Marine Corps stuff gets silly, he keeps me grounded. He never lets it bother him, and just says that it'll pass. It always does, and we keep rocking along.

MSgt Madeline's marriage to a former Marine illustrates how a spouse's prior service eases some of the potential problems described above:

> My husband was a Marine when I met him. His contract was up before mine and he got out. He's followed me around ever since. It's a huge role reversal. I'm the one that gets up at 5:30 a.m. to go to PT. I stand 24-hour duty at least once a month. He has to drop off and pick up the kids from school and practice and all. He has to pick up and move every three years. He has to find a new job. When our son came along, my husband decided that he wanted to stay home with him instead of putting him in daycare, so my husband was a stay-at-home dad. For the most part, he raised our son because I was focused on work. The good part of that was that I *could* focus wholly on being the best Marine I could be. I knew that my family was cared for. That, I believe, made all the difference in my career. I never had to worry about my family because my husband assumed the roles in our home that I couldn't at the time. It was about my career. I think because he was a prior Marine that he *understood* what was required. I don't think it was *easier* on him at the time. If anything, I think it might have been harder because of the role he had chosen to play. Not many men would do that. But, one of the reasons he did let his contract go was because we were ready to start a family and he didn't want us to be going separate ways, duty station-wise or deployment-wise. There's never the expectation, the knowledge that you'll be together as a

family. Your first priority is to the needs of the Marine Corps. He didn't want our children to have to experience that. He wanted us to be together to raise a family. So, he helped make it easy for me to be where I am in my career, but we've had to deal with personal issues like him not being able to find a job at a new duty station. But, we've survived this far. We've lived our marriage one duty station at a time. When we got to Parris Island and were told, "If you come here married, you'll leave divorced." Well, when we got married, we said that we were not going to be one of those military couples that didn't make it. Every three years we get a new duty station and we get a new start on our marriage. Going on 19 years of marriage, it's been three years at a time.

Similarly, SgtMaj Cadence enjoys her life with her husband, a former Navy enlisted man:

I have been blessed. He was in the military, so he was used to traveling around. We moved every three years and he'd pack up and go with me. He was an insurance salesman, so he was able to relocate and transfer, which made it easier. I do see that it's harder for a male civilian spouse to follow a female Marine than it is for a wife to follow her Marine husband. With the men today, these young guys being married to female Marines, I think their pride gets in the way a lot of times. They think they have to have the "better" job, make more money, be the provider, and that's difficult when you're moving around all the time. That male ego gets in the way of so much. Marriage is difficult enough, but when you add the Marine Corps, the challenges are incredible. It really takes special people to make these marriages last and so many don't make it. When we socialized doing Marine Corps things, my husband was always very supportive and would attend social functions with me and not feel uncomfortable. He's very proud of me. He'll be talking with a group of guys and they'll ask him something related to the Marine Corps and he'll say, "You need to ask her; *she's* the Sergeant Major." At first, they're shocked and confused. It's really funny.

Regarding the Marine Corps mating game, female Marines express their happiness at finding men that understand them and respect them for their strength and dedication. The sex and power differentials are mitigated or realigned to ensure all commitments and expectations are met for, hopefully, continued mutual satisfaction.

Marine-Marine Relationships

Often complicated to manage, involving frequent relocations and separations uncommon in the civilian world, marriage to another Marine manifests unique elements uncommon beyond the confines of the installation. Like any marriage, one key element is commitment. Like-mindedness definitely helps. Captain Maya describes why she believes it is difficult for female Marines to develop relationships with civilian males and expresses her happiness with her Marine spouse:

I haven't seen too many female Marine-male civilian relationships work out, like my female friends who are trying to date civilians…I'm very lucky because I have a husband who's a Marine, a grunt. I haven't seen too many relationships where the woman's a Marine and the man's a civilian work out. I can't think of a single one. I know they're out there, but the two that I knew of ended in divorce. I'm sure there's a little bit of angst when the husband comes home and has his wife going off to war and he's at home. I can't imagine that it doesn't make the man feel like a lesser man or a lesser person in at least some small way, either subconsciously or consciously. I'm not an expert on the human psyche, but I'm pretty sure that's part of it. Culturally, too, that's not the way it works. Even though we're starting to see more dual income parents in the workforce, I think we're still very much a society that has clear lines drawn between who is the primary breadwinner and who is the primary care-giver. I don't think that's necessarily "bad." I mean, I'm planning to get out and go into the reserves because I am the primary nurturer. But, I do think there are more women out there who are choosing not to have children or choosing or more typically "masculine" profession, whether it's business or the military or DEA Agent, you name it. I think more women are going in that direction, but it's still hard for a lot of men to accept that. I think you see women becoming "stronger" but not men becoming "weaker." I think "feminine" and "masculine" are more appropriate words. I think military, law enforcement, fire fighting…they're all masculine jobs, typically. So, it's hard for women in masculine-type jobs to find civilian counterparts that respect their strength…I am so thankful that my grunt loves and respects me for who I am, not what he or someone else expects me to be.

The data suggest that, for successful Marine-Marine relationships, gender negotiation is usually not a problem; however, competing commitments and time management are the priority and challenge. MSgt Darla relates how life as a Marine with a Marine spouse can get hectic:

> I took two classes the beginning of last year, and my husband deployed on me, which really ticked me off 'cause I was in the middle of my second-semester accounting class. So, here I was juggling three kids, college, a new dog and the house…Oh, my *God*. He was only gone for 60 days, but I really don't think I could have gone much longer! [Laughs] After that, I said that I am not going back to school for a while. The way it was planned out, we thought we had it organized, but the agency he works for changed it up on me, so I was like, "We're not doing this again." I made it through, but I've got a few more gray hairs. I did pass the class. My sister tells me they're hairs of wisdom. Of course, when it's time for him to go again, he'll go. When it's my time, I'll make sure we get a puppy before I leave. Bwaahahaha.

SSgt Tomasa also describes the challenges of being half of a dual Marine marriage:

> It's important to have a plan. When my husband goes to schools or deploys, it's no big deal. But, when I go anywhere, it's not easy at all. People say, "Well, you have your husband." I also have *four little girls* and this has to be well thought out. Just a simple thing like going to the rifle range, I have to check with my husband to make sure he doesn't have anything planned during that time and can assume all responsibility for the girls. Of course, he never asks me first. He'll just call like on a Thursday and say, "Hey, I'm going to the range next week." It's always just a given that his career is more important. I'm sure it never even occurs to him that he might need to check with me, but for me, everything has to be coordinated.

Because of the demands of the Marine Corps on a female's time and energy, being successful requires commitment, time management, and resolve. Having an understanding and supportive spouse can reduce the anxiety created by military service and its unrelenting demands. Conversely, being a married female Marine without the respect of her spouse can strain a marriage, especially if a female Marine feels like she is shouldering the emotional and social burdens of the household on her own. Children, obviously, only compound the situation.

Marine Corps Moms

Cpl Erica, a young woman married to another Marine, describes the difficulty of managing being a mom and a Marine:

> There is no "balance" in being a wife and a mother and a Marine. You just find the routine and you stick with it. Every day, you just find ways to make it all work. I drop my son off at day care, and I'm instantly a Marine NCO. I pick my son up from day care, and I'm back to being a mom. I pull into the grocery store, and I'm a wife stocking the pantry and the refrigerator, planning dinners, planning laundry. This weekend, I have duty, so I'm planning care for my son, making sure my uniform's ready, and all that. It never ends. When the baby's sick, I am the one that takes the time to care for him, not my husband. The leadership can see trends, though. Some Marines do have special needs children and SNCOs and all work with that, not just because it's mandated, but because it's the right thing to do to care for your Marines. They also see, though, the people abusing the system, and it's not always just the females.

In civilian life, the standard way for professional women to obtain the freedom to pursue a career is to hire other women to look after their children and clean their houses for them.[18] Of course, it goes without saying that women historically served the same functions for professional men, maintaining their home and caring for their children, but they received no compensation. Overlooked in van Creveld's statement is the fact that females historically have embraced kinship networks that include childcare responsibilities.

Women have rarely, if ever, been "kept" regardless of their race or socioeconomic status.[19] The contemporary need to pay for childcare is nothing more than the commercialization of a system that was previously unpaid labor performed in a kinship, barter atmosphere. Ironically, Charlotte Dowling lauds the American women's soccer team in the World Cup in the final game against China as "unencumbered and free and beautiful." Two pages previously, however, she gushes on their alleged freedom: "Not only can they handle the rigors of elite-level sport, they can manage their children while they're doing it…many taking their children along with them, *the nannies and grandmothers to help* as they travel around the world."[20] [Emphasis added.] Female Marines rarely, if ever, afford this luxury of nannies and camp followers as they traverse the globe with their weapon a rifle, not a soccer ball. Any feminist alleging women's emancipation from physical, social, and occupational constraints need look closer at the networks actually supporting that "freedom."

For female Marines, especially those intent on a career in the Marine Corps, management of their childbearing is a primary concern. Maj Lauren articulates the pregnancy stigma:

> If a female wants to have a family, I am good with that. When they come in and pump out a kid a year is when I start to have a problem with it. Some females still view pregnancy as a ticket out or as a paid vacation. PT ceases. Standing duty after a point ceases. If you don't like your shop, you move somewhere else. I've seen a few that have made pregnancy a career in the Marine Corps. They do just what they need to do to re-enlist and they know the system well enough so that they'll cry, "EO" if anyone tries to block that re-enlistment because of their "rightful decision" to raise a family or, in some cases, a litter. That's when it becomes sickening, and those are the ones that send the negative message out to male Marines. They're rare, but they're the ones that are remembered.

Fully aware of the pregnancy stigma, Captain Maya describes how her career was rocketing along until she married and experienced an unplanned pregnancy:

> I had an unusual career path because I got pregnant on my honeymoon. I cried and cried and cried and cried, because I thought, "Oh, my gosh, I'm slated to go be a Staff Platoon Commander [at Quantico]. I'm supposed to be a SPC [Series Platoon Commander] with Charlie Company, and they pick up in March, and…" I cried because we were planning on being DINKS, dual-income-no-kids, for awhile. I wasn't ready for it. I thought, "I'm gonna be a SPC at TBS", you know, like the best thing you can ever do is go back to TBS and be an instructor and instruct new lieutenants. And, oh, my gosh, I found out that I was pregnant, and…I didn't know what it really meant for my career. I thought everybody would hate me. I thought people were gonna see me as a sandbagger. Ummm, so I told my boss and I made him swear not to tell any-

body. And, I didn't tell anybody until I was, like, I was gonna *have* to start telling people because they would start to notice. There was another female instructor at TBS who had been pregnant and had a child and she didn't tell anybody either, so I just thought that would be the best way to handle it. But, I had a really good friend of mine, a male, who said, "You gotta tell people....First of all, it's great news; you should be happy. And, second of all, people are gonna wonder why you're not going to Charlie." And, that's what happened. People were wondering why I wasn't going to Charlie. I was slated for the Charlie Company staff and, all of a sudden, my name got pulled and they got this other guy. The staff had already started bonding and they were wondering why I wasn't going, and they were talking to their Company Commander, asking why [she] wasn't coming. And, nobody really knew. Rumors were starting. Not bad rumors, necessarily. And, I wanted to clear everything up, so I ended up telling people, and the exact opposite of what I feared would happen, happened. I work in a place where there are about thirty guys and, right now, there are three women. So, I instantly got thirty big brothers. They thought it was just great. At least, outwardly. It wasn't really necessarily from other captains that I got [insincerity]. I think some of the majors, the older instructors, weren't comfortable with it because I kept going to the field. I remember, it was probably May timeframe, and I would have been seven months pregnant, and I was out in the field in a flak jacket doing convoy exercises. And, one of the majors was really uncomfortable. I think he was concerned about me, for my personal welfare and the baby. He's a really good guy, and he was like, "I dunno...should you really be out here?"

In spite of her devastation at learning of her pregnancy, Maya continued to march in every capacity she was allowed even when males around her perceived that she was incapable or acting recklessly. Fundamentally, carrying that child in her body changed the dynamic of her status as a Marine Corps officer, but not in the negative ways she had feared it would. Similarly, CWO3 Leona describes the potential stigma that often accompanied pregnancy:

I do have to say I got a lot of kudos because I was still running PFTs seven months pregnant. Looking back, that probably wasn't the best thing, but that was what I felt I had to do to keep up and maintain that respect I had earned. I was driving five-tons while I was pregnant. Now, they won't even let a woman do that. Back then [in the early 90s], it was, like, "whatever." I do remember feeling like I had to do that to maintain my status in their [the males'] eyes. I would walk three or four miles when I was eight months pregnant even if I couldn't run, so that they knew I was out PT'ing.

SgtMaj Tiana describes her reasons for putting off children until late in her career at roughly her 16-year-in mark:

From the very beginning, my husband wanted kids but I told him that I wasn't sure if I wanted kids or not, so he never pressed the issue. Being a Marine is hard. When you're pregnant, there are a lot of things you can't participate in, that you can't do, and when you have the child, you don't throw yourself out there: "I'll go to the field. I'll go on the range. I'll go on deployment. Sure, I'll take your duty." You just can't do those things anymore. Your commitment doesn't stop, of course, but the level definitely changes. So many years later, we decided to have a baby and we had our first son in 2003. [Before that], I had a lot of things that I wanted to do, and I did them… I got to shoot on the Marine Corps Rifle Team and was distinguished. These are things that I just couldn't have done if I'd had a kid. I never would have gone to shoot. I mean, I could have shot there on the island [Hawaii] in division matches, but I could never have gone to Marine Corps matches because my husband was in Okinawa. There are so many things I would not have had the opportunity to do…well, I would have had the opportunity, but I would have had to decline if I'd had children. I don't regret a thing. I've always shot from the hip and been spontaneous in choices in my career, like being a Drill Instructor that first time. Things have just always worked out so well.

GySgt Claire describes being given a guilt trip when she told her SNCOIC that she and her husband were hoping to start a family:

When I was a young staff sergeant, I told my boss that my husband and I were trying to get pregnant. He was shocked and tried to make me feel bad about it, like I'd become a traitor. It is a difficult situation because we have to balance all of these parts of our life, and pregnancy and children definitely calls into question your commitment to the Marine Corps. No one can deny the fact that having children for a *female* Marine adds another complication to her career. Males just don't deal with this because the caregiver for that child is their wife. For us, we carry all the burdens.

Due to the pressures described above as well as others, many female Marines like Cpl Jewel elect not to have children at all while on active duty:

No children! I plan on staying in. I have so many things to do, but staying in the Marine Corps often hinges on weight issues for females. This is the first time in two years that I've been below my [weight] max...It's hard enough now with my frame and all of the muscle I carry, but as I get older, I am concerned that I will age like my mother. She's very active physically, but she struggles with her weight and her shape. That's another reason why I won't have children. If this is difficult for me now as a young woman to stay within the standard, what would it be like for me to try to lose baby weight? I can't even imagine attempting that. So far, I've been able to beat the tape, but I'm always borderline, and I hate that. A baby would destroy my career.

While many female Marines choose not to have children or plan their family around their careers, some just start a family and deal with the consequences. Sometimes it works out and sometimes it doesn't, especially for young enlisted female Marines whether or not they are married and single parents. They recognize the burden it places on the command as well as themselves. Cpl Lorna also admits to being a problem Marine, becoming pregnant early in her career:

When I got pregnant right at my one-year point [in Okinawa], they sent me back to the States because I was a single parent. I met my husband there; he's a corporal now and, of course, he stayed in Oki because we weren't married. My unit didn't want to deal with me having to move out in town 'cause I'd have to get a Japanese driver's license and all. Also, they just didn't want single females living out in town. I had a full year left on the island and I was only a PFC, so my pregnancy really did create a lot of problems... When I arrived in the States, my SNCOIC didn't judge me because I was a pregnant single female Marine. He just considered me another Marine that needed guidance, so he did what he could to help me get set up and ready for the baby, and kept me on track as a Marine. It was interesting, too, because after I checked in, we got another pregnant female just sent back from Okinawa, also. Since we were on the same track, she and I hit it off and became supportive of each other...My husband got back from Okinawa just in time for the baby's birth, but I'm almost like a single parent with him gone all the time. It's really hard. Financially, I wasn't ready at all for this. I had to get an apartment and a car, and then all the things that a baby needs. My husband wasn't ready either, so we're stuck in debt. Because the baby's so small, I am leaving him with my mom when I deploy [to Afghanistan] in a few months. As much as I'd like to leave him with my husband, I think that wouldn't work. No, not at all. We'll all be better off with my son at his grandma's while I'm gone. I could decline deployment for up to a year, but I want to go before he turns one, so he won't miss me as much.

Single Marines & Parenthood

SSgt India, a single parent with custody of her 16-year-old daughter, describes the challenges of juggling her roles as mother and Marine SNCO:

Male Marines have the best of both worlds because whenever they are chosen to be deployed or go on ship, or when they go to schools or MSG duty or to the Drill Field, or wherever, their main focus is on taking care of themselves. If they have a wife, all they have to do is pack their bags, if she doesn't do that for him, and make sure that their pay keeps going home to mama, their SGLI is up to date, they have a power of attorney and Will, give her the key volunteer information, and all that, and *poof* he's gone. But, for female Marines or single parents or dual-Marine spouses, the list is much more extensive and can be downright challenging. They deal with issues just beginning with: *Who* will care for my child? Do the children need to be sent home to grandparents or other relatives? How and when do I get the children there? What all do I have to set up to ensure their financial, medical, and educational care while I'm gone?

I have deployed twice so far. I was in Kuwait and then in Iraq. I'm getting to go to school Monday, and when I am running around and doing all of the things that I need to do, it's so much *more* than the males, other than the few male single parents I know. I can't just put my name on the list, pick up my ticket, and leave. I was telling my gunny the other day, "I have to make sure my daughter's medical, dental, and educational needs will be met. Because she's a teenager, it's even more challenging, because I have to disclose so much of our lifestyle to ensure her safety while I'm gone. My CO gets a list of her friends, her routes, her schedule, and all of the things that might get out of kilter while I'm away. My brother, who she's staying with, is a 24-year Marine, so he knows the system, but it will be challenging for him to now be the guardian of a 16-year-old girl. It's heavy on my mind that I can't make any mistakes in setting this up. There's no one to just step in and seamlessly take my place in all of the things I do for my family, my daughter, every single day. Marine wives are already taking care of their families, but I have to do all of that, too, while also being a Marine SNCO.

Validation of "Greedy Institutions"

As evidenced by these stories and so many others, mating and bearing children while on active duty dramatically alters a female Marine's relationship to the Marine Corps, supporting Buss' assertion that resource management for females can be extremely tricky.[21] Mating "well" can minimize some burdens, especially financially, but female Marines, particularly those with children, are often at the mercy of mounting occupational expectations, resource demands and limitations, and criticism for their decisions and situation. The adage used to be, "The Marine Corps didn't issue you a wife," and the Marine Corps was not graciously accepting of the outside demands placed on its young Marines with wives and families. Now, however, there are a multitude of family services in place to ensure the "quality of life" of married Marines and single Marines with children. Old habits and expectations, however, die hard and, while some Marines are accepting of this very natural process, others balk.

Military sociologist Charles Moskos asserted that in the institutional military, individual commitment and self-sacrifice is legitimated through the operation of normative values, which compel the individual to accept great demands on his [her] time and energy. The organization controls the demands: the individual does not choose when and how to comply.[22] The legitimacy for the institution to place its members at such physical risk is perhaps the greediest aspect of all.[23] Attempting to appease numerous greedy institutions creates the dilemma of trying to pack the challenges of "Marine-ing," "wife-ing," and "mother-ing" into a 24-hour day. These challenges weigh heavily on female Marines often because, unlike their male counterparts, they are often left solely responsible for many non-Marine Corps activities, socially and in the home. To address van Creveld's assertion of female privilege and put their role now as warrior in perspective: what makes superior human beings is the willingness to risk what we all, as human beings, value most: our own lives. Even if their lives are enigmatic, the elevated warfighting role of female Marines cannot be denied any longer. [24]

Female Marines attest that commitment, prioritization, time management, and a strong sense of self are the keys to survival and success in learning to appease numerous and competing greedy and unforgiving institutions within the military life they chose. Facing these demands can be daunting; surmounting them sometimes seems impossible. Thus, many female Marines find themselves trapped in situations where they must choose between the ones they love and the love they hold for the Marine Corps. The most difficult decision often is whether to leave the Marine Corps or stay.

Endnotes

1 van Creveld, *Men, Women and War*, 211.

2 Carol A. Kolmerten, *Women in Utopia: The Ideology of Gender in the American Owenite Communities* (Bloomington and Indianapolis: Indiana University Press, 1990): 73.

3 Buss, *Evolution of Desire*, 221.

4 Shields, "Sex Roles in the Military," 102.

5 Jennifer Mittelstadt, "'The Army is a Service, Not a Job': Unionization, Employment, and the Meaning of Military Service in the Late-Twentieth Century United States," *International Labor and Working-Class History* 80 (Fall 2011): 37-40.

6 Segal, "Greedy Institutions," 11.

7 Brownson, "Battle for Equivalency," 5.

8 Buss, *Evolution of Desire*, 30-32.

9 Brownson, "Battle for Equivalency," 2.

10 Mitchell, *Women in the Military*, 343.

11 Brownson, "Rejecting Patriarchy," 5.

12 Williams, *Gender Differences at* Work, 136.

13 Buss, *Evolution of Desire*, 27.

14 Paul V. Crosbie, "The Effects of Sex and Size on Status Ranking," *Social Psychology Quarterly* 42 no. 4 (December 1979): 341.

15 Buss, *Evolution of Desire*, 59.

16 Sjoberg, *Gender, War, & Conflict*, 32.

17 Moskos, "Toward a Postmodern Military," 23.

18 van Creveld, *Men, Women and War*, 213.

19 Catherine Clinton, *The Plantation Mistress* (New York: Random House, 1982), 20-21 & 162.

20 Dowling, *The Frailty Myth*, 222-224.

21 Buss, *Evolution of Desire*, 19-48.

22 Segal, "Greedy Institutions," 12.

23 Ibid., 16.

24 Christopher Coker, "Humanising Warfare, or Why Van Creveld May be Missing the 'Big Picture,'" *Millennium – Journal of International Studies* doi: 10.1177/03058298000290020201 (June 2000), 456.

Chapter Nine

Semper Fidelis

"*Semper Fidelis*" [Always Faithful] and "Once a Marine, always a Marine" mottos encapsulate the irreversible metamorphosis that occurs when one becomes a United States Marine. Thomas Ricks summarizes the experience succinctly:

> …the Marines are distinct even within the separate world of the U.S. military…Culture – that is, the values and assumptions that shape it members – is all the Marines have. It is what holds them together. They are the smallest of the U.S. military services, and in many ways the most interesting. Theirs is the richest culture: formalistic, insular, elitist, with a deep anchor in their own history and mythology.[1]

Always is a mighty long time. The tradition is rich and unwavering. Thus, recruitment and attrition are always a concern for personnel management on the macro-scale. On the micro-scale, similar to joining, leaving the Marine Corps can be difficult. Due to their small numbers, the departure of females from the military, similar to their arrival, is more pronounced than for males.[2] As previously stated, some American women become Marines on a whim or because they are running *to* or *away* from something. Pat Shields asserts that females are attracted to an institution, not an employer, a special institution offering unique benefits.[3] Of these benefits, the achievement of kinship and equivalency with males are but two of these benefits.[4] For example, Staff Sergeant Tomasa describes the bond and range of emotions indicative of both sexes in the Marine Corps and how they integrate, not separate, the male and female Marine experience:

> If I cry, it doesn't mean I'm weak. I'm just releasing pressure that a guy might release by punching a wall. Is he "weak" by doing that? Oh, hell, no. The difference, though, is that he's breaking something, and violence implies power. Tears aren't destructive so they're viewed as passive and weak, and that's just a *wrong*

way to look at these behaviors. The power of that emotion demands an outlet. I've seen plenty of male Marines cry and it sure as hell ain't because they're weak. It's because they *care*, and that's why I love them. We're all in this together, and we honestly love each other as brothers and sisters. This is where I will stay as long as they let me.[5]

The Marine Corps Culture in Transition

The contemporary Marine Corps straddles traditional, modern, postmodern, and postemotional philosophies, expectations, practices, and legislation. Sociologist Meštrović asserts that, as opposed to a postemotional civilian society, traditional or inner-directed societies always allowed enough privacy to the individual so that he or she could go through the motions of the traditions and rigid values yet maintain a reservoir of strictly private and authentically personal emotional reactions to these behaviors.[6] Although obviously going through postemotional changes, the Marine Corps remains a traditional society, a sub-culture, and one that encourages personal integrity and expression on an equal footing with group loyalty in appropriate measure. GySgt Tiffany describes her unit's and her response to the death of one of their teammates:

> My guys cried. When my MSgt was killed, I actually had a couple of guys walk over to me. I wasn't actually out on that mission, but I had a couple of guys walk up to me with tears in their eyes and I had to say, "You need to walk away 'cause I can't do this right now." I didn't cry in front of anybody. That was how I had to deal with it. When we lost Baker, I walked into the building and I said, "I am locking this door. Do not enter for five minutes." And, I sat down and bawled my eyes out. Then, I cleaned myself up and walked back out and we cleaned the HUMVEES.

The Marine Corps is a very public environment within its isolation from the external society. Emotion and passion are expected as part of the Marine's condition; emotion and passion motivate. Indulgence, however is not tolerated when it frivolously impinges on the effectiveness of the group. Each mistake, incident, or act of Fate, however, must be acknowledged and assessed, a lesson learned perhaps, and a commitment made to limit casualties in the future. Marines largely respect each other's need for privacy and "good"

Marines are never considered weak for the occasional expression of simply being human in extreme situations. Marine Corps "toughness" is the veneer for the incredible complexity within.

Kinship & Equivalency

In the U.S. Marine Corps, unrelated males and females from many geographic locations, socioeconomic backgrounds, ethnicities, and religions come together for the common goal of national defense and warfighting forging kinship bonds in the process. This endeavor, when females achieve equivalency with their male peers, instills in both sexes/genders a sense of trust, camaraderie, freedom, and power rarely experienced in civilian society. The tenets of evolutionary psychology identify innate sexual behavior patterns distinct to the two binary sexes, male and female; social constructs historically support biological imperatives. Theory and now these data underpin a very real framework in which the concepts of kinship and equivalency are critical components to understanding and mitigating the phenomenon of gender relations in the military.

Practically, if physical and professional respect and the resulting social harmony are to be achieved, and it must be in that order, both women and men must be recognized as linked together in a spiraling co-evolutionary process.[7] This is no "revolution" of woman against man as feminists allege. Rather, it is the best possible embodiment of connectedness and mutual esteem in the midst of a long trajectory of human existence and, obviously, success. To ignore or condemn our collective past is irresponsible. To forsake our collective potential future through dogmatic distraction and dictum is egregious.

The Marine Corps "Home"

Leaving this "home" of kinship and equivalency proves challenging for many female Marines, personally and socially. Female Marines should never experience shame for leaving military service voluntarily.[8] Similar to the withdrawal of Marines during the Battle of the Chosin Reservoir, they are not "retreating," they're just advancing in a different direction.[9] In reality, however, an opposite, negative impression of their choice to leave their Marine Corps often emerges.

For many female Marines, their experience in the Marine Corps is nothing more than a life stage just as it is for many male Marines.[10] For others, it is a vocation in the tradition of generations of professional warfighters.[11] Because males comprise 93% of the force, however, their comings and goings are not perceived as reflecting "deficiencies" in their male-ness. In contrast, however, a female leaving the Marine Corps is often perceived by herself and others as defeated by the experience due to her sex/gender. For whatever reason, she is unable or unwilling to remain on active duty.[12] They either "can't hack it," are "disciplinary problems," or they make "female issues" their priority in life instead of the Marine Corps.[13] This last reason is critical because there are likely very few male Marines who would be accused of such as excuse for exiting military service. If they did, they would likely be berated as "pussy whipped" rather than recognized as intelligent men making valid personal life choices. Again, regardless of the length or quality of their service or their reason for leaving, females often are marginalized because of their sex.

A Responsibility to Remain

When making the decision to stay on active duty or to separate, female Marines express a variety of compelling reasons for each option. The decision is always very personal and sometimes heartbreaking. For someone who has devoted so much of one's self to the Marine Corps, saying good-bye is simply not an option. The responsibility to their Marines and to the future of their Corps compels them to remain. SSgt Amber describes why she remains on active duty:

> I'm a Staff Noncommissioned Officer of *Marines*. Everything that I do is important. This isn't a job, this is my *life*. My leadership style is treating my Marines like my children. Not that I play down to them by any means, but I nurture them to become the best individuals and the best Marines that they can be. I want them to succeed. It's my responsibility as a SNCO to train the next generation of Marines, and if I haven't done every single thing that I can to prepare *them* to lead into the future, I have failed in my mission. The key is the continuity. Continue what works and change what doesn't. Marines improvise. It's what we do. It works. But, we can't cast aside our heritage because that's what brought us this far. Times change and, yes, the Marine Corps changes, too. But,

the essence remains, and that is our legacy for the future of the Marine Corps and for the United States of America. Honor. Courage. Commitment. *Semper Fi!*

Similarly, GySgt Pamela describes returning from years of Marine Security Guard (MSG) duty to a postemotional Marine Corps that she hardly recognized:

I had just been back in the Fleet for not too long and I was complaining that young Marines were just out of control. When I came in, even as a PFC, it was, "Yes, LCpl. No, LCpl." But, now I had returned to the Marine Corps, back in the FMF, and I was a sergeant and it was, like, "Yeah, I'll do it. No problem," from these kids! There was just no respect anymore. It seemed that it had changed that quickly, and a Marine at the brief stood up, and she said, "If you do nothing but complain about today's Marines and don't do anything to change or improve it, you're part of the problem. You're *not* part of the solution." I think that struck a chord with me. I was, like, "These damn Marines! These fucking lance corporals!" Blah, blah, blah. I remembered the respect that I had had for a LCpl as a PFC, and I was now a sergeant and I wasn't getting the respect I used to give two pay grades down! This is a personal challenge that I deal with daily, and it's my mission to adjust [correct] and support these young Marines. Tradition, pride, and success cannot be compromised. I am here to be that transition to the next generations.

Leaving in Disgrace

Whether the stimulus is positive or negative, many female Marines who remain on active duty describe the dutiful aspects of their service. For many, like GySgt Pamela and SSgt Amber, leaving the Marine Corps would be leaving a job unfinished. Others remain in anticipation of "what's next," anticipating other doors opening to new opportunities. For some, however, who choose to leave, the Marine Corps makes an indelible and sometimes ugly mark on the rest of their lives. World War II veteran Sandra describes witnessing a Woman Marine being "drummed out" of the Marine Corps while stationed at North Island, California, at the end of the war:

My First Sergeant was drummed out. She went to Tijuana and got drunk and went AWOL. They don't do that [drumming out] anymore, but it was quite a pageant. It was before the UCMJ. She stood in front of the whole group, was physically stripped of her chevrons, and then the drummer played while she marched away, reduced to nothingness in the Marine Corps. It was a terrible thing to witness. She stood there stoic, all alone, and she took it. She got on the bus to New York City. She didn't break down. She stood there and took her punishment, exactly like a Marine should, for what she knew was an unforgivable offense. This might not have been her first time [to defy Marine Corps expectations], but she sure paid the price for it and she did so with all of the dignity she could muster.

Serving forty years later, Cpl Lisa describes her Marine Corps experience and admits a deep sense of remorse for her decisions and action in the mid-1980s:

Basically, I left before they could kick me out. [Laughs] Really, I was like constantly on the verge of self-destruction. I had been struggling with alcohol abuse for the first couple years, well since right after boot camp, really. When my end-of-service came up, I'd had one alcohol related incident after another in my SRB and a DUI. They even put me in the psych ward at Camp Pendleton for two weeks. That was quite an experience! Even if I wanted to re-up [re-enlist], they probably wouldn't have let me. I don't blame them. The sad thing is that I haven't had a drinking problem since I got out and my therapist thinks that it was situational....I think that, maybe if I had received better guidance, I could have been a better Marine. I can't lay blame, though. I made my decisions and, as it was, I sucked. I was a drunk and I really didn't care about my responsibilities. I just went from one party to the next and barely did what I had to do not to get thrown in the brig, just barely, though. I sometimes wonder what I could have become if I hadn't been a drunk. [Sighs] It's like four years of my life that could have been really positive, but I totally wasted them. Yeah, veteran status with a Honorable Discharge looks okay on a resume, but it sure as hell wasn't honorable service by any Marine Corps standard. I'll regret that until the day I die.

Similarly, GySgt Iris describes her Marine Corps career as a roller coaster of pregnancy, weight control, bad paper, and ultimately, a Bad Conduct Discharge (BCD):

I had a corporal working for me and we got to be really good friends. She was dating a Marine who was into drugs. He was a pill popper and they were doing a lot of Xanax. Please bear with me 'cause that is kind of a preface for where this story goes. So, I go and take a PFT. I weigh in and I'm eight pounds over. I get a 2nd Class PFT. Now, had I gotten a 1st Class PFT, the weight would not have been an issue. My husband was coming off the Drill Field at this same time, and we had orders to Yuma, Arizona. Now, though, I have to go see the Base SgtMaj, and he tells me that he's not going to PCS me because I am over my weight, but that he is PCS'ing my husband. Here I am, a gunny with five kids and Iraq in full swing. We both have orders to deploying units in Yuma, but I am getting stuck at Parris Island. Oh, and prior to this, I had gotten NJP'ed as a Gunnery Sergeant for financial irresponsibility. I was pregnant at the time I got NJP'ed, so things had gone from bad to worse. So, me in my infinite wisdom, and it was the dumbest thing I ever did, I get with this corporal. I call her up and I say, "Okay, I know how to lose some weight real quick," and I bought a half-ounce of cocaine on a Monday. This was the Monday after the Friday they told that they were not going to PCS me with my husband. My intent was to use it for a week, not eat at all, weigh in, and get the hell off of Parris Island. My husband, who never did drugs before he joined the Marine Corps, did the cocaine with me. That night, they busted the guys, the Marines dealing. They had been surveilling [*sic*] the Marines that were doing this for *two years*. The day that I bought the cocaine was the day that NCIS went in, and they busted a male corporal…it was a corporal from our battalion who I didn't even know. He worked in our admin, and was running the cocaine up from Miami. So, NCIS showed up at my house at 1:30 in the morning, searched our house, but did not find the cocaine. They found a baggie with residue, though, so they made me take a urinalysis. They made me, my husband, and the female corporal from my shop take urinalysis. We all pop positive. I was court-marshaled the same day as my husband, and I went to the brig for six months. I ended up serving four months. I was demoted to private. I was kicked out with a Bad Conduct Discharge, and they ended up giving me a Special Court Marshal. They charged me with distribution to my husband. So, basically, I was charged with the same thing as the drug dealers. That's why I got six months. I would have retired this year. It would be 20 years in October, and I would have retired as a Gunny because I had the NJP. My husband and I are still together, but as civilians now. Not a day goes by that we don't miss the Corps and regret the way we left.

The female Marines interviewed who left the Marine Corps under duress express feelings like they let everyone down, especially themselves. All affirm that they wish they could go back and make different choices. No former female Marines interviewed who left under disreputable circumstances place blame solely on the Marine Corps for the bad choices they made; they accept responsibility for their decisions and actions.

Relinquishing One Obligation in lieu of other Obligations

Again in support of Buss' assertion that females face particular difficulties in securing resources on their own, especially while raising children, and actively choose male mates to provide for them and their children, many females leave the Marine Corps in deference to their husband's career and in acceptance of their roles as wife and mother.[14] Although not likely a reason that males leave the Marine Corps, females making these choices support evolutionary theory as well as social theories regarding mating behaviors of the two sexes.[15] For example, Cpl Whitney describes her decision to leave at her EAS [End of Service]:

> My EAS is next year, and I won't re-enlist. I am going to be a full-time mom and go to college. My husband, though, will re-enlist. That's another reason why I won't re-enlist. If he's going to have to go through all this every day, it's not right for all of us, as a family, to have both of us on active duty. Deployments are too much right now. I can't ever allow both of us to be deployed at the same time. So, I'm leaving.

Captain Christine, too, is leaving a stellar career to resume her role as nurturer:

> I don't think you can both be a wife and mother *and* deploy to war. 100% is only 100%. If you give 90% to your profession, your family's getting 10%. I think that's one of the big myths of my generation, thinking that we can have it all. Well, we *can* have it all. I just don't think we can have it all, all at once. So, that's why I tell my girls [junior female Marines], "Go and have it all in the Marine Corps right now and see if that's really what you want because, later on when you start having a family, you're gonna have

to make some tough choices. Maybe the choice won't be to stay home, but to put your kid in daycare for 11 hours a day, but you're gonna have to make those tough choices. Well, as long as you're a Marine, the Marine Corps is gonna get theirs." And, that's the way it *has* to be. I don't say that in a negative light. When you have people's lives on the line, that's what it takes. But, you only have 100%, and that's it. The Marine Corps is all-consuming and that's why we're so good.

Major Devon, after 21 years of service working her way up from private, has decided to leave because "it's time." She takes time to laud the "wholeness" of female Marines she sees every day:

You look at 4th Battalion today, though, and you see female Drill Instructors who are hard as nails and incredible leaders, but then you see them in town and they're hot. If you met one at the gym or the grocery store or at church in Beaufort, you'd never know she was a Drill Instructor. They're beautiful, feminine women. They have families and talk about waxing and their nails. [Laughs] They manage to balance all of the demands on their personality. They're tough when they need to be and soft when they choose to be. It's a beautiful thing to see, and I like to count myself as one of those well-balanced lady Marines. [Laughs] I'm pretty sure my husband and my sons would agree! I play football with them *and* bake cookies. I compete in triathlons *and* I like to sew. It's just who I am, and I'm pleased with it. Coming up on retirement, I joke about having to learn how to be a girl even more. I think if I leave the Marine Corps now, I'll be too hard or too over-the-top for most civilians to be comfortable with. However, I am a prosecuting attorney, so I could blame the "hard edge" on that "masculine" lawyer status. [Laughs]

Similar to the stories included in the previous chapter, the female Marines interviewed chose or will choose to leave the Marine Corps to assume a more total responsibility for their families. Although they express their love of the Marine Corps and appreciate the experiences they have had, they also admit that it is too difficult to "be all things to all people" and they believe that each role suffers when each requires 100% of themselves. Some simply want to relinquish the masculine aspects of themselves. In the words of SSgt Lynnann:

As I get older, the feminine side of me demands more expression. I need to totally chill. Plus, I don't want to be perceived as "that old female Gunny" hanging on past her years of being awesome. Faltering after success would be detrimental here in the Marine Corps. So, it's time to leave now while I am strong and lean, but just starting to feel the wear in my joints. I'm sad about it, but I think this is the best decision for everyone and the organization.

Readjusting to Civilian Life

Former female Marines describe difficulties returning to the civilian community whether it was after four or twenty-four of service. They assert that they miss the Marine Corps, especially the camaraderie, the kinship, they felt within the military environment that does not exist in the "real world." Some express out-right disgust at the absence of standards and high-quality leadership. Iris continues her story, describing the challenges being a civilian after 16+ years on active duty, including a successful tour on the Drill Field:

> Coming back home and being a civilian and trying to re-adjust to being a civilian and the differences....Civilians are harder to deal with than Marines are. You just have such a work ethic as a Marine. They told you what you had to do; you just got it done, no questions asked. There were no excuses. The mentality is so different than in civilian life. I think that's what the hardest thing is. I work with a bunch of guys and I'm the secretary, hanging onto the bottom rung of the organization. Although I do so much more than that, it's really hard to look at them and think, "I used to lead Marines." I have done more in my life than they'll ever do in theirs and most of them are in their 50s. And, to know the leadership…I have a branch manager and he's just the nicest guy in the world, but he's a piece of shit. I mean, he's never at work. Unlike the peons, he takes about seven, eight weeks of vacation and gets paid for it. It's so opposite of what the Marine Corps was because you took care of your Marines first. The civilian world is just so backwards. Taking care of your own and leadership by example should be the priorities…I really miss that.

Former SSgt Francine, too, finds civilian life a pale comparison to her life as a Marine:

> I'd have to say that my favorite thing about the Corps is the brutal honesty. Having been an enlisted Marine from private through

Staff Sergeant, and now as an executive with a Ph.D., I find that the way people pussy-foot around language and situations in the civilian world is insane. At any level of my Marine experience, there was always someone waiting, sometimes eagerly, to tell me I was screwing up, but there also were situations in which I could truly shine and receive accolades. There was no requisite politeness that went with the corrections I received whether from a man or woman, black or white, officer or enlisted. There was never the sense that someone would not correct me because I would cry "lawsuit" or "discrimination," and I never remember anyone feeling so threatened by me that they would ever withhold saying, "Hey, nice job," say, on the rifle range, a PFT, or whatever. There really was a kind of mentality that promoted honesty and self-growth. You knew where you were in the rank structure and what was expected of you at that level. Women, of course, always had higher standards set for them, though. [Laughs] Sometimes, I hate my job and being a civilian, in general, because I can't simply say what I think for fear of repercussions. I feel like I am kissing peoples' asses when I have to modify, "soften" if you will, a rejection letter because someone might cry or feel inadequate when they read it. Hey, if I am sending you a rejection letter, you *need* to feel inadequate and figure out what measures you need to take to *become* adequate! Other than that, I can't help you. [Laughs] I *hate* being surrounded by a bunch of over-sensitive slackers and whiners!

Others express that, as female veterans, civilians cannot "classify" them as such. Her status as a female Marine Corps veteran proved troublesome for Captain Karen in the LAX airport. As a former linguist, she translated her military skills into a lucrative civilian occupation with an international corporation and was on her way to Japan to conduct a series of executive seminars. She describes what transpired:

I was waiting at the bar for my flight and sipping Jack and Coke because, even after all those hours in the air in the military, I absolutely *hate* to fly. I was chatting with a couple of fellow travelers, both men about my age, about really nothing at all when a bunch of Marines coming back from somewhere apparently landed and they descended on the bar like a plague of locusts. In their Charlie uniforms and radiating that swagger that only Marines radiate, the bar came alive. The two guys I'd been talking to, and others, were shaking hands and buying drinks for these guys. It was really pretty cool, electric, even. Then, one of the young officers standing in the midst of us mentioned that he was going home to Houston to be an OSO, and just conversationally I popped off, "Oh, that's awesome, I originally enlisted at MEPS in Houston and my OSO was in College Station." The group fell silent and stared at me. I don't know if the frame of reference didn't exist for them because I had no uniform on. Maybe it was because I was a woman? Perhaps, there was a disconnect because of my comparative age. All I know is that I felt like a nasty bug in the midst of these amazing, beautiful butterflies. No one asked; I just disappeared into the background. I *never* felt that way in the Marine Corps, and I never want to feel that way again as a veteran.

Former female Marines express that their transition back to the civilian world was more difficult than the transition into the Marine Corps. They attribute this to leaving the civilian lifestyle in search of lofty ideals, which they found in the Corps.[16] A few former female Marines describe elevated empowerment, achievement, and success as veterans as a result of their service. Other female veterans exit the passionate military environment to an alien, bland civilian world. They express disappointment in what they perceive as a return to laxity and chaos. Meštrović succinctly encapsulates their recognition of postemotional society:

> All of the primal passions discussed from Aristotle to Hume to the present become shadows of their former selves. Anger becomes indignation. Envy – in the form of the traditional covetousness of a neighbor's cow, children, wife, whatever – now becomes an objectless craving for something better. Hate is transformed into a subtle malice that is hidden in all sorts of intellectualizations. Heartfelt joy is now the bland happiness represented by the 'Happy Meal.' Loving really becomes liking. Sorrow, as the manifestation of affliction, anguish, grief, pain, remorse, trials, tribulations, and sadness, is magically transformed by the TV journalist's question, 'How do you feel?'[17]

Female Marines, experiencing the transition from the Marine Corps back to civilian life, again identify as enigmas. Their spheres and their competency within them never quite correlate to those established by society. They always exist separately and, often, very much alone.

Keys to Success: Sage Advice
from the Trenches

This project utilizes their personal stories to illustrate the mosaic of the female Marine experience, an experience enigmatic to the uninitiated.[18] The gamut of emotions, expectations, and behaviors indicates that no single experience is the only possible experience. Rather, each female must negotiate her own challenges to achieve whatever relative success or experience she desires. It should be clear by now, however, that management of her *very self* and her female-ness, is paramount to success in the Marine Corps. Although mostly absent in the early AVF generation, today's female Marines are privy to an abundance of wisdom and the mentorship of senior females, enlisted and commissioned. When asked what advice they would offer a young female to be successful as a Marine, their responses were consistent. Captain Maya recalls a story from OCS, and passes on advice that her platoon sergeant gave her in the summer of 1999:

> She said . . . "Close your eyes and picture the perfect Marine. He's about 5'11" chiseled chin, straight back, small waist, right? High-and-tight?" And, I'm like, 'Yes, Platoon Sergeant." And, she replies, "Okay, well, of course, that's everyone's vision of a perfect Marine. You will never fit that stereotype. So, just stop trying. Don't ever try to be that. Just be yourself. Remember that eyes will be on you because you are a small population in the officer corps." . . . [Ignore] the rumors about being promiscuous, or bitchy, or a lesbian, she said; just affect what you can. Don't worry about that. Be your own person, but you are going to have to be the best person because everyone's going to be looking at you. Where a guy might be able to blend in, there's no blending in when you're 5% of the population.

SgtMaj Cadence adds that a female must establish herself and make decisions with the future in mind:

> If a woman wants to make a career of the Marine Corps, she must establish herself first. As a PFC or a Lance Corporal, you're very young. I don't have anything against women having babies, but there's a lot expected of you at those lower ranks. I made the conscious decision based upon my goals not to have a child before I was at least a sergeant. If you're going to make it a career, you

have to plan. It's hard in those first years, and that's the time to establish who you are and where you want to end up. This is not a place to get rich. If you stay in only four years and you get out with children, you better have yourself squared away because there will be no more financial support for those dependents. Your civilian employer isn't going to give you a housing allowance. You better be prepared because being young isn't an excuse when you've had education about personal and financial responsibility. I encourage all of my young Marines, male and female, to think hard about the decisions they make while they're in. This is a great time to establish yourself. Then, think about marriage and babies and all that. You've got plenty of time. Work on *you* first.

GySgt Claire agrees that a young female entering the Marine Corps must have herself together:

The key to success for the young female Marines is that they need to know who they are when they join the Marine Corps. If they don't, it will be too easy to get lost if they're trying to create an identity in this environment of stereotypes. You have the pretty girls, the fast girls, the butch girls, and the new ones coming in have to figure out where they belong in the mix. Then you have the females that already have themselves together and they're the ones that are the best Marines. They're not worried about who they are and how to be a Marine. They know who they are and they can focus on *being* a Marine.

CWO4 Renae adds the perspective of the company one keeps in the Marine Corps. The sex ratio and the purpose of the organization are not going to change, so a young female must be up to the situation:

The key to success for females in the Marine Corps is to keep your head about you. Just do what you're taught by your Drill Instructors and do what you believe is right. If you have a doubt, don't do it. It's not rocket science. It's not hard. Morally, making the right decisions will keep you on track across the board. If you respect yourself and you respect what you're doing, you'll be fine. You have to be hard, though. It's a hard environment. It's about warfighting and you're in the company of warfighters. If you can't deal with that, go Air Force or Navy. This isn't the place for people who constantly second-guess morality or fairness. It has to be pretty damn cut-and-dried in your head when you go outside that wire, so you need to be prepared for that. Not everyone can do this. That's why we've always been the few and the proud. We don't care about bigger numbers. Numbers just provide fodder. We care about heart and character and commitment.

LtCol Abigail asserts that personal responsibility is critical:

The standards [in the Marine Corps] are there to live by, and I welcome that. Sure, I could get laid any time I wanted, but that wasn't who I was. My career and my personal goals were more important than sex. Believe me, sex is the easy part. Developing a reputation as a competent and trustworthy Marine was what I was all about…The key is personal and professional responsibility. It's particularly tough for that young group, the 18- to 24-year-olds far away from home. A person has to understand where she, *or he* because it's true for both sexes, fits into the bigger picture of the Marine Corps. The Marine Corps has a proud tradition of honor and integrity to uphold. If a Marine is gonna get snot-slinging drunk and act like an ass out in town, that reflects negatively on all of us. If a female Marine's gonna participate in group sex at the BEQ with a hired photographer, *that* reflects negatively on all of us, too, and even more so because she's breaking more taboos than the "typical drunk Marine." I like to say, "If you wouldn't do it in front of your mama, don't do it."

Captain Kai offers no-nonsense advice to young women considering the Marine Corps:

I think the most important thing that any female thinking of joining the Marine Corps needs to know is that she needs to be able to measure up. No excuses. No bullshit about, "I'm a woman; I'm being discriminated against." She needs to be prepared to meet the standard and she needs to have a thick skin and handle herself confidently and competently. This sounds horrible, but women with self-esteem issues and eating disorders and all of those things that our screwed up culture produces do *not*, I say again, do *not* need to join the Marine Corps because they can't handle it. They cannot handle it and that's not fair to anyone. They will fail as leaders, and they will be eaten alive because warfighting is a dog-eat-dog operation. You have to be running swift and deadly with the winning pack, the Devil Dogs. You have to be able to hold your head up in their company without assistance. Period.

SSgt Manda acknowledges her status as a competent Marine leader who also revels in her female sexuality. She offers advice to make a difference in the real day-to-day of the Marine Corps:

> My presence here reminds them [the males] of being admonished by their mother as they were growing up, challenged and goaded by their sister, and adored by their girlfriend all at the same time. I think this mix of emotions that truly amazing female Marines inspire is the key to a successful 21st Century integrated Marine Corps. To my guys, I am not abusive of them as a sexual being. I don't flirt and tease. I certainly won't be pigeonholed into a maternal role by them. However, I am totally sexually female in my presence, in my physicality and the ways I interact with them. I can't escape that. We live together; they see me, hear me, and smell me. I am part of the clan, but I am still a mystery because, as a female, I represent so many compelling, competing, and complementary emotions in them that they bring into the Marine Corps. We all do. I think we, as females, have powers over men that other men will *never* have, and I think they really hate that, but they love it at the same time and that scares them. Maybe that's where all the misogyny bullshit came from? It's not right to be manipulative, but if we learn and master how to positively use our female assets without making ourselves needy and only sexual, I think we'll honestly have the best of all worlds. I think that is where our "power" as female Marines truly lies.

Semper Fidelis

By their own admission, female Marines struggle balancing the demands of the Marine Corps with the demands of self and family. Many separate from active service to accommodate their familial and other personal responsibilities. This confirms the premise that female Marines do tend to leave for "female" reasons. Females continue to report placing their husband's career before their own, their unwillingness to deploy and leave their children, and their frustration in trying to give 100% of themselves to multiple demanding consumers of their time and energy as primary reasons for leaving military service.[19] Ultimately, strong women, Marines after all, often choose to relinquish the strength and independence achieved in the Marine Corps, physically, financially, and socially, in lieu of assuming the more traditional roles of wife and mother. These choices, however, should not be condemned. In the words of Matt Ridley:

No moral conclusions of any kind can be drawn from evolution. The asymmetry in prenatal sexual investment between the genders [sexes!] is a fact of life, not a moral outrage. It is "natural."... That something is natural does not make it right [or wrong]... Prejudice, hate, violence, cruelty—all are more or less part of our nature, and all can be effectively countered by the right kind of nurture. Nature is not inflexible but malleable. Moreover, the most natural thing of all about evolution is that some natures will be pitted against others. Evolution does not lead to Utopia.[20]

The Marine lifestyle demands sequestering many physical and personal aspects of one's self, and many females struggle with striking an appropriate balance between their sexuality/femininity in an environment that expects the display of masculine or neuter characteristics.[21] Females labeled as weak or sexually manipulative most clearly illustrate this challenge. With an abundance of female leadership and literature now present, young females have the opportunity and support structure to perform well, provided that they arrive at Parris Island or Quantico with the requisite physical abilities and mental attitude.

The experiences presented here confirm that Marine camaraderie is not just a "man thing." Rather, it is a "Marine thing." Those who earn the title accept certain responsibilities that go along with it. The stories presented also describe ways in which females continue to deal with negative stereotypes and struggle to maintain positive impressions of their Marine professionalism in relation to their personal lives. Regardless of the validity of their reasons for doing so, however, females' motives for separating from the Marine Corps are generally questioned. They are viewed as uncommitted, not up to the challenge, or selfish in rejecting the Marine Corps in deference to their roles as women. This is unfortunate because many, many female Marines, not unlike their male counterparts, also remain *Semper Fidelis.*

Endnotes

1   Ricks, *Making the Corps*, 19.

2   Shields, "Sex Roles in the Military," 111.

3   Ibid., 106.

4   Brownson, "Battle for Equivalency," 3.

5   Partial quote previously used in Brownson, "Rejecting Patriarchy," 4.

6   Meštrović, *Postemotional Society*, 66.

7   Brownson, "Battle for Equivalency," 20.

8   Max Lerner "The Shame of the Profession," in *War, Morality, and the Military Profession* Malham M. Wakin, ed., (Boulder, CO: Westview Press, 1986): 139.

9   Oliver Prince Smith, "Retreat of the 20,000" *Time Magazine,* December 18, 1950.

10 Shields, "Sex Roles in the Military," 106.

11 Ibid., 102.

12 Ibid., 105.

13 Ibid., 109.

14 Buss, *Evolution of Desire*, 25.

15 Shields, "Sex Roles in the Military," 109.

16 Ibid., 108.

17 Meštrović, *Postemotional Society*, 62.

18 Brownson, "Battle for Equivalency," 4-5.

19 Shields, "Sex Roles in the Military," 109.

20 Ridley, *Red Queen*, 180-181.

21 Brownson, "Battle for Equivalency," 16-20.

Chapter Ten

Future Research & Conclusions:
Hail & Farewell

This final chapter *pauses* in sharing the experiences of females in the Marine Corps, the qualitative research data captured in my interviews. Although greatly condensed herein, these stories unabashedly illustrate a complex mosaic of past and contemporary experiences in the military microcosm, that of women and United States Marines. Using a transdisciplinary approach, my purpose was to frame, analyze, and interpret these experiences within established theoretical tenets of evolutionary psychology and sociology.

The thesis of this research proposed that the on-going "problem" of women in the military persists because of the unrealistic expectation that females are "equal" to males. Ultimately, I believe I have I delivered fully on my objectives. I propose and the data support that, instead of feminists' conceptions and expectations of female "equality," the pragmatic ideals of kinship and equivalency actually facilitate women's success in the U.S. Marine Corps. Grounded in a sense of kinship emerging from an extreme and shared experience, equivalency encourages empathy, camaraderie, and trust between and among contributing community members.[1]

Specifically, historical and contemporary literature on "gender relations" in the military inadequately address the crux of the phenomenon due to its complexity on the macro-level (e.g., institutional, societal) as well as the micro-level (e.g., interpersonal and small group relationships).[2] Daily, however, social science theorists and researchers around the globe in a multitude of disciplines replicate ineffectual theories, hypotheses, and methods on this subject. They construct surveys, count heads and "yes's" or "no's," and tabulate Lickert scale values. They run regression analyses into concise presentations of sterile, impotent data. Academics present papers, chair presentation and posters sessions, and receive awards and accolades. They ponder the intersection of human nature, history, ethics, and social behavior, afraid to really tackle the hard questions while the phenomenon continues all around us constantly, incrementally evolving. This is but one reason why climate surveys and investigative reports cannot meaningfully address military and warfighter dynamics; they provide an often inherently biased snapshot of opinion limited by time and space with no capacity for longitudinal integrity. Many are fundamentally flawed simply by poor design. At worst, they are institutionally manipulated.

In contrast, however, my qualitative data innovatively inform the literature affirming that, to date, "gender relations in the military" research has decisively missed the mark. Specifically, these data confirm that antiquated theoretical concepts of patriarchy,[3] misogyny, and tokenism must be wholly discarded or, at the very least, intently scrutinized and re-tooled, if appropriate, to salvage theoretical/conceptual elements relevant to contemporary realty and discourse.

From analysis of the data this work represents, I assert that limited-discipline and stringently dogmatic approaches employed to investigate and explain the phenomenon of sex/gender relations issues in the U.S. military are egregiously inadequate. In contrast, I propose that an innovative, pragmatic, and historically supported theoretical framework based upon the concept of "equivalency," rather than presumed "equality," between the sexes be adopted. This conceptualization of relationships acknowledges sex/gender disparities and ensures appropriate investigation and analysis of related sexual, social, and occupation equity in the military.[4] It eliminates, categorically, the presumed patriarch's benevolent condescension and the radical feminist's revolutionary anger in lieu of a pragmatic, reciprocal, and co-evolutionary approach to understanding this phenomenon.

Warfighter Solidarity

Of greater importance and based upon the data herein, the male-female disparity issues should no longer be identified as "gender relations" in the military. Instead, the phenomenon has evolved into a holistic conception of a non-gendered *warfigher solidarity*. Although it seems that the title "U.S. Marine" would encompass it, this concept emerges from a systems approach grounded in a practical framework of equitable and pervasive standards and expectations for both sexes encompassing personal, professional, and occupational elements. Thus, it acknowledges and integrates the actual complexity of the military experience on multiple levels and is applicable across all military services.

Fundamentally, acknowledgement of equivalency between and among peers fosters empathy. Empathy, in turn, facilitates camaraderie. Camaraderie, in turn, facilitates small unit cohesion and, thereby, potential operational success from the bottom up. Thus, warfighter solidarity should adequately reflect relationships of perceived competency and trust regardless of sex/gender in the U.S. Marine Corps. Females as well as males establish themselves proficiently within this system to be successful. Equitable standards for all promote institutional success. This is not gendered; this is reality. Further, these data are *not* generalizable to all women simply because women should no longer be lumped together as one "thing" even though, in reality, all military men are. Radical feminists' assumption of all women's potential for success in the military is ludicrous.

My data confirm that full acceptance of women as equivalent Marines (i.e., professional and occupational subordinates, peers, and superiors) is a reality, but requires relearning by both sexes to dispel previously learned male-female role expectations. Some women and men who attempt to break out of these fixed roles must exhibit substantial fortitude as they risk rejection and failure in their attempt to discover and develop themselves along unchartered courses.[5] Generally, all Marines must resist pressure to embrace antiquated, biased expectations of females. In turn, females must strive to meet or exceed mandated standards and expectations to ensure their own acceptance and success.[6] Further, they must not use their sex to differentiate themselves from males to garner favor in their work relationships and the enactment of their military duties.

Within the big picture of biological and social evolution, the stereotyping and the stigmatization of women in the Marine Corps are based partly on fact and partly upon individual behavior. They are too often used out of convenience or ignorance when applied to the entire group, which these data support. I challenge King's assertion that stereotypes are utilized as a "cultural resource for men."[7]

I counter, again, that they are only as powerful as the group, females as well as males, allow. The numbers are irrelevant; conviction sways. Specifically, when Marine Corps' core values, traits, and principles are embodied, women have a greater likelihood of inspiring their peers and earning their respect as equivalents.[8]

These data further confirm that within each Marine, male and female, there exists various degrees of hardness of heart and depths of compassion, of selfishness and altruism. The relationship is one of kinship, at times adversarial, yes,[9] but ultimately co-evolutionary. Buss asserts:

> Although conflict between the sexes is pervasive, it is not inevitable. There are conditions that minimize conflict and produce harmony between the sexes. Knowledge of our evolved sexual strategies gives us tremendous power to better our own lives by choosing actions and contexts that activate some strategies and deactivate others. Indeed, understanding sexual strategies, including the cues that trigger them, is one step toward the reduction of conflict between men and women.[10]

The Postemotional Marine Corps

The generational and individual voices of female Marines emancipated herein represent transformative changes in the Marine Corps reflective of corresponding societal changes. Anticipating this transformation, preeminent military sociologist Charles Moskos stated:

> The hallmark of the Modern military was that of an institution legitimated in terms of values and norms based on a purpose transcending individual self-interest in favor of a presumed higher good. Members of the American military were often seen as following a calling captured in words like "duty, honor, and country." The Postmodern model, however, implies much more. The structure, makeup, and purpose of the armed forces changes as well as its values. The basic point is that a Postmodern military ultimately derives from the decline in the level of threat to a nation and, in the American case, certainly, the rise in identity politics based on ethnicity, gender, and sexual orientation.[11]

In making this statement in pre-9/11 America, Moskos did not anticipate the rise in threat to national security nor the corresponding response of ethnic, gendered, and sexual individuals to rise to the occasion. Specifically herein are the voice of females of many ethnicities and backgrounds who strive for equivalency with male Marines as women, citizens, and guardians of the American way of life.[12] Although to the uninitiated it may appear to be a large, lumbering bureaucracy, the Marine Corps is not a static institution. The image of the robotic Marine or the toy soldier is wholly false. It is the diverse array of Marines within the structure who breathe life into it. Females play an integral part in this life-giving process. Only comparatively late did society begin to recognize that just as human biological life needs the two genders to sustain it so, too, social life needs the interaction of masculine and feminine values to sustain its institutions.[13]

The data indicate that, in the Marine Corps, one learns to see beyond color and religion and sex if one's comrades prove themselves to be Marines first and foremost.[14] Similarly, Meštrović asserts that postemotional humans try desperately to recapture the emotional energy that used to be achieved only through collective effervescence, yet fail more often than they succeed.[15] In the Marine Corps, the collective effervescence, the *esprit de corps*, although perhaps weakened by the postemotional American society from which it draws its members, is alive and well.[16] It also remains accessible only to those young men and women willing to sacrifice themselves to ensure the success of the organization and the mission. Not all succeed in this framework of kinship and equivalency, which is true of male as well as female Marines. W. E. B. DuBois astutely summarizes leadership qualification: "…authority means the recognition of the fact that all cannot lead because all are not fit to lead, but we must listen to the noblest not to the loudest…"[17] In the Marine Corps, this is an achievable status regardless of sex/gender. It cannot be purchased. It is not secondhand, living a life as flotsam carried by forces not of our making, a sort of second-hand living.[18] It is earned. It is a coming together of human individuals, collectively, at a level incomprehensible by civilians. The media, try as they might, simply cannot capture this spirit; it remains enigmatic.

Female Self-management: Biology, Sexual Subjugation, & Warfighting Solidarity

Between birth and death humans exist as they have for eons as sexual and social beings, hoping beyond their conscious thought to be immortalized through their progeny. Historically, only a few heroes, after all, are remembered beyond their own generation or their own clan.[19] Men and women, military or non-, throughout history are united in this endeavor of reproduction.[20] In this, all are kin and equivalent. Christine Williams asserts that "[m]en's combined economic *and emotional* self-interests in perpetuating gender differences will ensure their persistence as long as men monopolize the dominant positions in the military."[21] [Emphasis in the original.] One must recall and acknowledge, however, Max Weber's assertion that the subjugated in large part choose to facilitate their subjugation.[22] In contrast to radical feminist battle cries most recently espoused by G.L.A. Harris,[23] Buss argues that human history validates the assertion that victorious revolutions of the oppressed over their oppressors represent the exception rather than the rule. Practically, men cannot be united in their interests of oppression, even in principle, for the simple reason that men are primarily in competition with other men. In American culture, men do not desire to oppress all women, for they have sisters, mothers, daughters, and nieces whom they respect and desire to protect, defend, and cherish.[24]

Female Biology & Power

One of the most important findings of this research is that female Marines simply cannot escape their biology and, further, have no desire to do so. Female Marines do not desire to be "honorary men" in the Marine Corps or in American society.[25] All women interviewed expressed great pride in being sexually female and socially women, and enact decisions to express their female-ness throughout their military and personal lives. The vast majority of female Marines interviewed possess the intelligence, prowess, and physical skills *at least* equivalent to those of their male counterparts. They expect no more from their males peers than recognition of that fact and the respect their contribution deserves. Collectively, they acknowledge and abhor the limited, often inferior status and resulting limited power access for women in American society. They assert that inclusion in the combat arms and the corresponding require-ment of Selective Service for women are not solutions. Most feel unable to alter the situation except as individuals enacting personal agency through their direct relationships with others. Their personal solutions were to seek the best of both worlds, which many found in the Marine Corps.[26]

In the Marine Corps, to achieve equivalency, female Marines must master and manage their biology, their very female-ness. Through physical fitness, they master their physicality. They must prove that they are physically strong enough, fast enough, robust enough, to compete with their male peers and earn their trust and respect. In the field, they master the bodily functions of urination, defecation, and especially the abhorrent menstruation to ensure that those "nasty" aspects of being human do not interfere with the mission, logistically as well as aesthetically. In stressful situations, they must master their emotions as their male counterparts must. All other things being equal, it is their sex that is the only divisive ele-ment. Thus, through management of the very things that define her physically as "female" or socially as "feminine," a female Marine can truly excel in the Marine Corps.[27] The data support that there are occasions when the distinction as "female" doesn't matter so much, and a female Marine can relax and enjoy being a woman in a man's world with whatever roles and results that entails.[28] However, she must always be on her guard to meet the physical, occupational/leadership, and social expectations of those around her because she knows whatever elusive ground she has gained towards equivalence with her male counterparts can be easily lost, often not even by her own doing, but as a result of the behavior of other females.[29]

Sexual Subjugation

The data herein imply that sexual harassment, assault, and rape in the military, or anywhere else, will never be eradicated until females can protect their own sexuality against aggressors.[30] With sexual freedom comes personal responsibility. This begins with es-tablishing boundaries in relationships and diligence in maintaining them.[31] Rape is, indeed, a crime. Based on their experiences, female Marines involved in this research believe that false accusation of rape should be a crime as well. Additional research and findings on this topic are forthcoming.

Female Solidarity

Contrary to the expectations of radical feminists,[32] female Marines are their own worst enemy in their diversity of purpose, commitment, and actions. Even in 1949, Simone de Beauvoir recognized that women are more closely bonded with their male "oppressors" through kinship and mating than to other women.[33] In sum, if mating is the female priority, warfighting is not. Meštrović asserts that the heritage of Durkheim's view of sociology should make one aware that the individual is powerless to change the course of social events without the help of his or her fellows through collective effervescence. And, it must be a *spontaneous, genuine collective effervescence*, not deliberately rehearsed and planned public spectacles.[34] [Emphasis added] Female effervescence captured in these stories clearly indicates diversity among women and alignment with male Marines and the mission rather than sex/gender solidarity.

At a Crossroads of Understanding

Not surprisingly, then, there remains a distance between female Marines as "sisters" in a shared experience. There also exists tension between female Marines and their civilian counterparts regarding the perception of female Marines' status and commitment as women and warriors. They remain an enigma; a chasm of incomprehension and misunderstanding that must be bridged. LCpl Sadie offers a solution:

> I think that a lot of civilians think that women shouldn't even be in the military. What that does is close off so many opportunities for other women who really *want* to be here, whether for the opportunity to become a leader or to earn money for college, or so many other reasons. When they [civilians] say, "You can't do this because you're a girl," there are so many trickle-down effects of that single statement. No, men and women are not the *same*, but we're equally important and we should all be valued as such. Females in the military set the example of what women can be when they don't allow society to keep them on the shelf. It's hard, yeah, but it's extremely rewarding. To earn a title and the ensuing respect is worth so much more than being handed stuff. If you're just taking what a man offers you without contributing to the big picture yourself, you're part of the problem.

SgtMaj Tiana chimes in as well regarding the solidarity between male and female Marines recognizing that each has a role to play:

> Marines are trained as riflemen from the time they get to recruit training. We are all taught the same basic things, and we don't expect to be treated any differently. You can't compare us to other services. We're wholly different. Unless a person knows Marines, they shouldn't take someone else's opinions or imaginings of us and think it's truth. Female Marines are just as hard working as male Marines. We're in those deployment rotations just as much. If you don't see us on CNN, we're not hiding. You're probably seeing the 93% of the Marine Corps, which is male, and we, the 7% are working our asses off, too, just not on screen. If we're not in the trenches with the combat arms, it's because that is not our place to be because the American people, the President, and the Commandant don't want us there. But, we're out there doing the convoys, doing the security points, doing the road repairs, doing the IED detonations. You think we're not in danger? Think again. We are in harm's way, but we're not waving a banner. We're too busy doing our jobs as Marines.

LtCol Abigail describes the changing location on the performance bell curve of females in the Marine Corps and the recognition of their contributions to the organization:

> Female Marines, historically, were viewed as the outliers on the Marine Corps bell curve. They were either outstanding or they sucked. This was the result, I think, of *looking at the women* as an anomaly just by their presence. Today, though, I think we're able to look at women as part of the larger group instead of as that smaller, enigmatic microcosm of the Marine Corps, and we're seeing more and more females clustering in the middle right there with the majority of the males. If we look at pink [female] and blue [male] dots, I think most people would be surprised by where the pink dots are in the larger mix. It's contrary to our traditional way of thinking about it, so maybe it's a good starting point for real discussion about male-female performance [in the military].

LCpl Alexandria, 21 years old and two years in the Corps, displays great wisdom in her youth:

> The only thing that I see as a problem is that some people are

intimidated by Marines, in general, and with females, in particular. They can't even wrap their mind around us, conceptually. It's really all about balance, though. A measure of toughness and a measure of kindness. A bit of hard core and a bit of flexibility. The Marine Corps actually values all of these qualities and so many more that seem to be at odds with one another, but it's all about choosing to take the appropriate action in the appropriate circumstance. Just read the Marine Corps Code of Conduct and the leadership traits and principles. That's not a "man's" code. That's a code for *living*.

The Marine Corps fosters an environment of loyalty and respect, both of which are earned traits, and not dispensed frivolously by Marines. The opportunity to engage in relationships based upon mutual admiration exists for males and females in the Marine Corps. Although more challenging for females to achieve because of the physical challenges and the stereotypes they face, those who achieve parity with their fellow male Marines truly and rightly appreciate their accomplishment. This builds trust and respect: equivalence. SSgt Manda describes an experience on the verge of Desert Shield/ Desert Storm that exemplifies the potential beauty of relationships between Marines, male *and* female:

One summer in the early 90s, I was the Admin Chief at a professional military school in San Diego. Everyone was running around getting ready for an IG inspection one morning after PT. This was right before Desert Shield really kicked off and everyone was on edge about that. The message traffic from CMC was heavy and constant and growing more and more ominous each day. We knew something was coming, but no one knew what would really transpire or how we'd really deal with it when it did. Add to that the inspection, and the entire School was on a hair-trigger. So, before the inspection, I was already in my Charlies and good-to-go and had time to make a run to the Comm Center for an urgent pick up. I returned to the office with a Top Secret that the XO, who was actually an Amtraker, needed to see ASAP. His door was closed, so I knocked and announced myself. And, from inside, he yells, kinda growls really, "I'm in my damn skivvies but, if you think you can handle that, Sergeant, you're welcome to come in." I had to laugh a bit at that to myself, but it just never occurred to me *not* to go in. So, I did. The major was standing at his ironing board, ironing his trousers. He had his Corfams [shoes] on and, of course, the traditional garters holding his socks up and his Charlie shirt tail down over his skivvies. He was a Viet Nam and Korean War vet and his chest was just stacked with ribbons, and, at that moment, to me, he looked like the *perfect* Marine. *Perfectly* squared away, but for the missing trousers, of course. He was the absolute essence of masculinity while standing there at an ironing board. Salt-and-pepper flat top, broad shoulders, trim waist, long strong runner's legs. An absolutely *perfect* male with a Sunbeam iron in his hand. I handed him the message and waited while he read it in case I needed to prepare an urgent response or distribute or whatever. There was never a moment of discomfort between us, only sheer professionalism and mutual respect. It was an awesome experience, really, to be so aware of his physicality and have it feel so pure, you know, the *exactly right way* for male and female Marines to feel about each other, physically and emotionally. So, he looked me in the eye and sighed. With a trace of a smile, he said, "We'll worry about this one later, I think." I nodded, and turned to leave. Then, he said, "Please secure that hatch on your way out, Sergeant. I don't want *all* the gals to see me naked," and he smiled. I smiled back. "Aye, sir," and we continued to march toward Iraq, the *first* time. Whenever I have a bad day, I think of that moment, and I realize again why I am here, and why I stay. It's for him and all the Marines like him, before and after.

Many female Marines relate similar stories about points in their careers when they felt a similar purity of emotion and physicality with their male counterparts. Those moments make the trying times a bit easier to bear. In the opinion of female Marines, basking in mutual respect with your fellow Marines, to *be* a Marine, is *true* women's liberation.

Closing Remarks

Samuel P. Huntington argued that the military officer corps alone shoulders the responsibility for national security. The service member's responsibility is the military security of her client, society. The discharge of the responsibility requires mastery and skill; mastery and skill entail acceptance of the responsibility.[35] Further, Huntington asserts that the professional military officer must apply technical knowledge in a human context.[36] Rather than the officer corps exclusively bearing this weight, these data indicate that Marines, male and female, at all levels of the chain of command embody the persona of the "strategic corporal." The acts of every member of the military have direct impact on local hearts and minds. A single mis-step by any member can receive immediate exposure on 24/7 news programs, with the potential for significant impact on public opin-ion.[37] Again, these data support the awareness by female Marines of their responsibility in the military as well as indicate a great capac-ity for competence as fully sanctioned warriors should that role be imposed upon them. In the military, the distinction between insiders and outsiders is more important than distinctions between the sexes. In overcoming negative stereotypes through competency, individu-als are judged by how effectively they perform tasks and are prized for their contribution.[38] The key is developing kinship and achieving equivalency.

This research provides the missing scenes from all Marine Corps movies ever made and books published.[39] It is, perhaps, the most important one. It is about *all* Marines, male and female, and reflects the society from which they come and that they swear an oath to protect against all enemies, foreign and domestic. Marines will tell you, as they have here, that their experience is not all brava-do and shock-and-awe. Sometimes, it's achingly boring. It's about trusting and covering each other. It's about mourning lost or ruined buddies and living with that constant gnawing, that wondering if there was that one thing you could have done differently to modify the outcome and bring them home with you whole. It's about bring-ing up the next generation with the traditions of the past but with an eye to the future. It's always about awareness of one's place in the big picture while tending to one's immediate responsibilities. It's about feeling excitement for the opportunities and demands the next rank, duty station, or deployment. It's about saying good-bye bravely.

The myths of the hyper-masculine, misogynist military and the bitch-slut binary must be eradicated. Females in the contemporary military are victims only when they allow themselves to be. The American public *must* be made aware of the reality, the intensity of the military environment and the unique challenges female Marines face as well as the competencies they do or do not possess. Instead of an enigma, female Marines should be acknowledged and lauded for their contributions as women and as warriors. Female equivalency with male Marines is an achievable status, but it takes some-one of extraordinary character and the commitment to self-sacrifice. In the sage words of EOD GySgt Tiffany: "This isn't reality TV. We're young and we're fit, but it's a real possibility here, and when it comes to dying, we all end up equal."

Endnotes

1   Brownson, "Battle for Equivalency," 3-4.

2   Ibid., 4-6.

3   Brownson, "Rejecting Patriarchy," 6.

4   Brownson, "Battle for Equivalency," 20.

5   Schwartz and Rago, "Beyond Tokenism," 74.

6   Brownson, "Rejecting Patriarchy," 2-4.

7   King, "Women Warriors," 3.

8   Brownson, "Rejecting Patriarchy," 5.

9   Sharon Erickson Nepstad, "The Continuing Relevance of Coser's Theory of Conflict," *Sociological Forum* 20, no. 2 (June 2005): 336.

10 Buss, *Evolution of Desire*, 13-14.

11 Moskos, "Toward a Postmodern Military," 27.

12 Brownson, "Battle for Equivalency," 3.

13 Coker, "Humanising Warfare," 451.

14 Brownson, "Battle for Equivalency," 5.

15 Meštrović, *Postemotional Society*, 102.

16 Brownson, "Rejecting Patriarchy," 2-3.

17 W. E. B. Du Bois, "The Spirit of Modern Europe," in *Social Theory: The Multicultural & Classic Readings*, ed. Charles Lemert (Boulder, CO: Westview Press, 1993), 186.

18 John J. McDermott, *The Drama of Possibility: Experience as Philosophy of Culture* (Fordham University Press, 2007): 273.

19 S. Brock Blomberg, Gregory D. Hess, and Yaron Raviv, "Where Have all the Heroes Gone?: A Rational-choice Perspective on Heroism," *Public Choice* 141 (2009): 511.

20 Ridley, *Red Queen*, 64-65.

21 Williams, *Gender Differences at Work*, 138.

22 Brownson, "Rejecting Patriarchy," 2.

23 Harris, *Living Legends and Full Agency*, 297.

24 Buss, *The Murderer Next Door*: 108. Note: Buss' statement regarding the "desire to protect and defend" addresses a long-held cultural belief until now never scientifically elucidated that females should not be allowed in combat because males would choose to protect them rather than focus on warfighting. From an evolutionary psychological perspective, the reason men desire to protect women is because women carry the future of humankind in their wombs. Thus, the females' ability to reproduce must be protected at all costs and the biological imperative is that males protect it.

25 King, "Women Warriors," 4.

26 Brownson, "Battle for Equivalency," 6.

27 Ibid., 8-9.

28 Brownson, "Rejecting Patriarchy," 5.

29 Brownson, "Battle for Equivalency," 14-16.

30 Ibid., 17-19.

31 Ibid., 19-20.

32 Ibid., 13.

33 Simone de Beauvoir, "Woman as Other," in *Social Theory: The Multicultural & Classic Readings*, ed. Charles Lemert (Boulder, CO: Westview Press, 1993), 369.

34 Meštrović, *Postemotional Society*, 147.

35 Samuel P. Huntington, "Officership as a Profession," in *War, Morality, and the Military Profession*, ed. Malham M. Wakin, (Boulder: Westview Press, 1986): 30-33.

36 Ibid., 31.

37 Nix, "American Civil-Military Relations," 94.

38 Shields, "Sex Roles in the Military," 108.

39 Sjoberg, *Gender, War, & Conflict*, 25.

Appendix A

Marine Corps Rank Structure

Enlisted		
E-1 Private (Pvt)		
E-2 Private First Class (PFC)		
E-3 Lance Corporal (LCpl)		
E-4 Corporal (Cpl)	}	Non-Commissioned Officers (NCOs)
E-5 Sergeant (Sgt)		
E-6 Staff Sergeant (SSgt)		Staff Non-Commissioned Officers (SNCOs)
E-7 Gunnery Sergeant (GySgt)		
E-8 Master Sergeant (MSgt) - technical	}	
E-8 First Sergeant (1stSgt) - administrative		
E-9 Master Gunnery Sergeant (MGySgt) - technical		
E-9 Sergeant Major (SgtMaj) - administrative		
E-9 Sergeant Major of the Marine Corps (SgtMajMC)		
Warrant Officers		
W-1 Warrant Officer (WO)		
W-2 Chief Warrant Officer 2 (CWO2)		
W-3 Chief Warrant Officer 3 (CWO3)		
W-4 Chief Warrant Officer 4 (CWO4)		
W-5 Chief Warrant Officer 5 (CWO5)		
Officers		
O-1 Second Lieutenant (2ndLt)		
O-2 First Lieutenant (1stLt)		
O-3 Captain (Capt)		
O-4 Major (Maj)	}	Field-grade Officers
O-5 Lieutenant Colonel (LtCol)		
O-6 Colonel (Col)		
O-7 Brigadier General (BrigGen)	}	Flag Officers
O-8 Major General (MajGen)		
O-9 Lieutenant General (LtGen)		
O-10 General (Gen)		

Appendix B
The Articles of the Code of Conduct

ARTICLE I: I am an American, fighting in the forces, which guard my country and our way of life. I am prepared to give my life in their defense.

ARTICLE II: I will never surrender of my own free will. If in command, I will never surrender the members of my command while they still have the means to resist.

ARTICLE III: If I am captured, I will continue to resist by all means available. I will make every effort to escape and to aid others to escape. I will accept neither parole nor special favors from the enemy.

ARTICLE IV: If I become a prisoner of war, I will keep faith with my fellow prisoners. I will give no information nor take part in any action, which might be harmful to my comrades. If I am senior, I will take command. If not, I will obey lawful orders of those appointed over me and will back them in every way.

ARTICLE V: When questioned, should I become a prisoner of war, I am required to give name, rank, service number, and date of birth. I will evade answering further questions to the utmost of my ability. I will make no oral or written statements disloyal to my country or its allies or harmful to their cause.

ARTICLE VI: I will never forget that I am an American, fighting for freedom, responsible for my actions, and dedicated to the principles, which made my country free. I will trust in my God and in the UNITED STATES OF AMERICA.

Appendix C

Marine Corps Leadership Traits

JUSTICE	JUDGMENT
DEPENDABILITY	INITIATIVE
DECISIVENESS	TACT
INTEGRITY	ENTHUSIASM
BEARING	UNSELFISHNESS
COURAGE	KNOWLEDGE
LOYALTY	ENDURANCE

Marine Corps Leadership Principles

- Know yourself and seek self-improvement.
- Be technically and tactically proficient.
- Know your Marines and look out for their welfare.
- Keep your Marines informed.
- Set the example.
- Ensure the task is understood, supervised, and accomplished.
- Train your Marines as a team.
- Make sound and timely decisions.
- Develop a sense of responsibility among your subordinates.
- Employ your command in accordance with its capabilities.
- Seek responsibility and take responsibility for your actions.

Glossary of Selected Terms and Acronyms

Due to the jargon and acronyms often used in the Marine Corps, a glossary is provided to explain briefly terms not generally present in the civilian vernacular.

4th Battalion: The battalion at MCRD Parris Island, South Carolina, that trains all Marine Corps female recruits; formerly WRTC (Women's Recruit Training Command)

8th & I: Marine Barracks Washington, DC, is the oldest active post in the Marine Corps, and is located on the corners of 8th & I Streets in southeast Washington, D.C. Marines at 8th and I perform ceremonial and security missions.

[SNCO] Advanced Course: PME [Professional Military Education]; a course for Marine Corps gunnery sergeants (E-7s) to further develop their skills as Marine leaders

All-Volunteer Force (AVF): Initiated in 1973, the All-Volunteer Force (AVF) ended conscription [in practice, not in policy] and became the U.S.'s first attempt to maintain a standing voluntary military.

Alphas: The Marine Corps Service "A" uniform, the most formal service uniform combination, consists of forest green and khaki elements and a coat and tie are always worn.

Avionics: An enlisted MOS in which Marines install and repair aircraft electrical systems, including weapons systems

Barracks: Living quarters for unmarried or unaccompanied Marines. In initial training, the barracks are open squadbays. After initial training, barracks are usually hotel- or apartment-style structures.

BAS (Battalion Aid Station): Battalion-level medical location where Navy Corpsmen conduct sick call, treat Marines for non-critical medical issues, and triage more serious conditions to the nearest Naval Hospital

Battalion Commander: The highest ranking officer in a battalion, usually a lieutenant colonel or colonel

BCD (Bad Conduct Discharge): Separation from service under "other than honorable conditions" and as a result of a general or special court-martial

BEQ (Bachelor Enlisted Quarters): Barracks where unmarried or unaccompanied enlisted Marines live in garrison

Body bearers: Six body bearers ceremoniously carry the casket of a deceased Marine to his or her gravesite and at the conclusion of the funeral service tri-fold the American flag and present it to the Marine's next-of-kin. One of the ceremonial details/drill companies at 8th and I that performs to crowds attending parades and ceremonies in Washington, D.C.

BOQ (Bachelor Officers' Quarters): Barracks where unmarried or unaccompanied Marine officers live in garrison

Blood stripe: Marines' bravery was rewarded with the blood stripe to honor the blood shed by Marines who stormed Chapultepec Castle in Mexico City on September 13, 1847. The red stripe on the Dress Blues trousers now represents all Marines who have given the ultimate sacrifice and is worn by all NCOs and SNCOs, beginning with the rank of Corporal (E-4), and the entire officer corps.

Boot Camp: Initial training of Marine recruits located at MCRD [Marine Corps Recruit Depot] Parris Island, South Carolina, for all female recruits and male recruits who enter the Marine Corps from East of the Mississippi River. MCRD San Diego is the Marine recruit training site for male recruits enlisting in the Marines Corps from locations West of the Mississippi River. Each has its own unique topographical and psychological aspects. Both are argued by their graduates to be the "best"/most challenging location for creating U.S. Marines.

Bosses' Night: Once a month, the ranks have the opportunity to mingle at one of the clubs on base and "bosses," typically NCOs and SNCOs, buy alcoholic beverages for their junior Marines and engage in casual conversation, play games (e.g., Spades, shoot pool, darts, etc.) that they would not normally engage in because of UCMJ prohibitions against fraternization.

Box kicker: An enlisted Marine in the Supply MOS 30xx, Supply Administration & Operations

Break ranks: To leave a formation, sometimes authorized (e.g., road guard) and sometimes not (e.g., a Marine faints and others "break ranks" to assist)

Bulkhead: A naval term to describe what civilians call a "wall"

Bulldog: Officer Candidate training/testing at Quantico, Virginia, for NROTC, Marine Option, college students; shorter in duration but not intensity compared to the standard ten-week OCS course

Butter bar: A casual term for a Second Lieutenant (O-1); also known as a "Boot Louie"

Cammie make-up kit: A compact with a mirror similar to a woman's cosmetics compact, but containing camouflage colors such as white, light green, loam, and sand

Camp followers: People who hang around the military, usually to provide goods or services that the military does not provide. One example is prostitutes, but the term is also often stretched to include dependents and military groupies who attempt to marry into the military to secure its benefits.

Candidate (or Officer Candidate): The designation of an individual in initial training (Officer Candidate School) seeking to become part of the officer corps of the Marine Corps. A baccalaureate degree is required or eminent during training (e.g., Bulldog), and the service requirements are above and beyond those of enlisted Marines.

Cattle car: There are various types, but basically an enclosed cargo-type trailer with a bus-type door on the side to haul people around. Sometimes it has windows and benches to sit on along the side, but not always. Marines are typically packed in like cattle, hence the name.

CAX (Combined Arms Exercise): A highly advanced war game conducted at MCB Twenty-nine Palms, California, utilizing Marine Air Ground Task Force (MAGTF) training and weapons

Chain of command: The hierarchy of Marines in one's organization. For example, for a boot private, an engineer, the chain of command is:

Fire Team Leader, Squad Leader, Platoon Sergeant, Company First Sergeant, and Battalion Sergeant Major. Officers are also in his/her chain of command, but perform different functions in their officer billets. For this example, they would be: Platoon Commander, Company Commander, and Battalion Commander. See Appendix A for the Marine Corps rank structure.

Change of command: A ceremony in which the authority and responsibility of high-level billets are passed from an out-going Marine leader to an in-coming Marine leader, and may be held for senior enlisted as well as senior officers. Usually a reception follows for VIPs.

Charlies: The Marine Corps Service "C" uniform, the least formal service uniform combination, consists of forest green trousers or skirt and khaki short-sleeved shirt and never a tie

Checking in: When arriving at a new duty station, a Marine must "check in" with numerous individuals/logistics within his/her command and on the base, such as billeting/housing, the armory, administration (S-1), etc.

Chevrons: Indicates an enlisted Marine's rank, worn on the sleeves of service uniforms and the collars of utility uniforms

Chow hall: The building where Marines not on commuted rations or receiving Basic Allowance for Subsistence (BAS) eat for free

CMC (Commandant of the Marine Corps): A four-star general and the highest ranking officer in the United States Marine Corps and a member of the Joint Chiefs of Staff who reports directly to the Secretary of the Navy

Combat Fitness Test (CFT): Three-event, scored physical fitness test comprised of a timed boots-and-utes sprint over 880 yards, counted lifting of a 30-pound ammo can over the head for two minutes, and a paired (with another like-sized Marine) 300-yard simulated maneuver-under-fire course. Scoring is different by sex and graduated by age.

Command and Staff College (CSC): Per their mission statement, "CSC educates and trains its joint, multinational, and interagency professionals in order to produce skilled warfighting leaders able to overcome diverse 21st Century security challenges."

Company Commander: Usually a senior First Lieutenant or Captain who exercises command and control over the unit and may exercise non-judicial punishment (NJP) authority over the personnel in the unit

Confidence course: A multi-station, cross-country obstacle course designed to challenge a Marine physically as well as mentally in order to develop confidence in his/her physical abilities

Corporal's Course: Required PME; per their mission statement, this course, "provides the Marine Corporal with the education and leadership skills necessary to lead Marines. The Program of Instruction places emphasis on leadership foundations and a working knowledge of general military subjects."

Cover: Worn on one's head, civilians call this a "hat." Another usage is defined as artificial or natural protection from enemy fire, such as seeking "cover" behind a tree or a vehicle or a Marine providing "cover fire" for a buddy exposing him/herself to the enemy; hence the phrase, "I've got you covered."

Crossed rifles: The depiction of two rifles crossed over each other exists on the rank insignia, chevrons, of enlisted Marines from the ranks of Lance Corporal (E-3) through Master Sergeant (E-8) between the stripes (above) and the "rockers" (if any) below.

[The] Crucible: A grueling 54-hour field exercise in which recruits hike for miles, roughly 40, and participate in night infiltration courses and team-building scenarios, all on eight hours of sleep (in two four-hour naps) and with 2.5 MREs that they must ration throughout the training evolution. Recruits learn to lead *and* to follow, *both* critical skills for United States Marines throughout their careers. After mastering the Crucible, recruits have earned the Eagle, Globe, and Anchor emblem, and are called, "Marine," for the first time.

DACOWITS: Established in 1951, the Defense Department Advisory Committee on Women in the Service is composed of women and men appointed by the Secretary of Defense. Their "mission" is to assess and develop recommendations relating to women in the military and now military families as well. Although many appointees have never served on active duty in any of the services, DACOWITS' recommendations have historically affected or dictated laws and policies pertaining to military women.

Dark Green Marine: A descriptive term used to distinguish African American Marines from non-African American Marines; usage of this term is no longer acceptable

Deck: A naval term used to describe what civilians call a "floor." It also describes what civilians call "floors" of a building, such as "the top deck" being the highest floor of a multi-floor building.

Delayed Entry Program (DEP): A program that allows time from signing a contract to reporting to recruit training for up to one year

Deployment: Transferring under orders with one's own unit or to leave one's home command to another command to engage in training operations at sea or abroad, or to engage in warfighting in a "hot zone"

DoD (Department of Defense): The Federal department responsible for all activities related to national defense and the U.S. military

Dog and Pony Show: An event orchestrated to impress VIPs; usually involves inspections, parades, or other "show of readiness" activities

DOR (Drop on Request): An option in Officer Candidate School (OCS), but not in recruit training, by which a candidate may choose to leave the initial training environment and return to civilian status

Dress Blues: The equivalent of "black tie and tails" for civilians. As with the service uniforms, there are various "levels" of Dress Blues, including the "modified Blues" often worn by Marine recruiters.

Drill Instructor: An enlisted Marine, male for male recruits and female for female recruits, charged with training recruits during initial training; usually an elite Marine who exhibits the highest standards of physical performance, occupational competence, and the highest integrity

Duty: A 24-hour watch stood at one's work space, barracks, shop, or any one of a variety of other locations on base, on ship, or in the field. A Marine standing duty may also be called "the Duty."

EAS (End of Active Service): Conclusion of a Marine's active duty contract. A military contract usually includes required reserve time as well, but the EAS concludes the 24/7/365 commitment of "active duty."

ECP (Enlisted Commissioning Program): A program by which an enlisted Marine, active or active reserve, who possess a baccalaureate degree may pursue, via OCS, a commission as a second lieutenant in the United States Marine Corps. Some restrictions apply.

Elliot's Beach: The location on MCRD Parris Island, South Carolina, that field exercise and weapons training occur

Enlisted ranks: Private (E-1) has no rank insignia and is referred to as a "slick sleeve," progressing through the level of E-9, the highest of which is the Sergeant Major of the Marine Corps. See Appendix A for the USMC rank structure.

EO (Equal Opportunity): MCO P5354.1D w/Ch 1 Marine Corps Equal Opportunity (EO) Manual (Short Title: EOM) of 14 April 03 summarizes within its 75 pages, "Unlawful discriminatory practices within the Marine Corps are counterproductive and unacceptable. Discrimination undermines morale, reduces combat readiness, and prevents maximum utilization and development of the Marine Corps' most vital assets, its "people". The policy of the Marine Corps is to provide equality of treatment and the opportunity for all Marines to achieve their full potential based solely on individual merit, fitness, and ability."

EO Advisor (EOA): This billet has its own Marine Corps Order, MCO 5354.3B of 17 Jun 02, a document of only nine pages, which includes in the Commanding Officer's Screening/Interview Guide a succinct description of duties. Of note are, "while incidents of race, gender, and religious inequities still occur, the EOA's mission is not one of advocacy for any particular group. Rather, their focus is readiness and mission accomplishment. Their efforts promote teamwork and understanding among Marines."

EOD (Explosive Ordnance Disposal): Among the many activities performed by Marines in MOS 2336, EOD Marines locate, disassemble, destroy, and document all manner of ordnance to include IEDs (Improvised Explosive Devices), which are a particular and effective threat to Marines currently serving in the Middle East.

Expeditionary Warfare School (EWS): A lengthy residence course for Captains (O-3) at Quantico, Virginia, that provides career-level professional military education with emphasis on combined arms operations, warfighting skills, tactical decision-making and Marine Air

Ground Task Forces in amphibious operations, preparing its graduates to become commanders and staff officers within the FMF.

FAP (Fleet Assistance Program): Not to be confused with the Financial Assistance Program (FAP) governed by MCO 7220.43B or the Marine Corps Family Advocacy Program (FAP) governed by MCO P1752.3B, the Fleet Assistance Program (FAP), in existence much longer and governed by MCO 1000.8 of 12 Jul 1994, allows for Marines to be FAP'ed to another location on their base to fulfill duties other than in their designated MOS, such as working at the pool, the Base stables, or a multitude of other locations. Although often considered a dumping ground for "problem" Marines, pragmatically the FAP program perhaps fulfills the needs of two commanders simultaneously, removing a "problem" for one while providing a "warm body" for another.

Fam Fire: "Familiarization firing" the M16A2 was the mandated training for female Marines in the eras when female Marines were expected to not be "too masculine," roughly from the Free a Man to Fight generation, when they were offered the opportunity to hold the weapon, until female Marines were required to qualify as riflemen in the mid-1980s. From the AVF generation prior to the qualification requirement of all Marines, WM's SRBs stated "fam fire" and the date since no score was kept.

[The] Field: Living away from garrison for training purposes. In the field, obviously, many of the "luxuries" of being in garrison are absent.

Field day: a massive group cleaning of any area that needs (or doesn't need) to be cleaned, including barracks, shops, outdoor areas, etc. Usually performed Marine Corps-wide every week on Thursday afternoons.

Fire and Forget: An association made to a weapons system in which a chip is programmed to hit a designated target regardless of the movement of that target. Once identified, a target will be destroyed because it cannot evade its predisposition as a target. Also referred to in a leadership scenario where a leader disseminates "fire and forget orders," meaning that the individual does not need to follow up because the mission as dictated will be accomplished without further intervention.

Fleet Marine Force (FMF) and also called "the Fleet": After initial training and assigned schools, the FMF is the "real" Marine Corps where a

Marine lives every day of his/her contract whether in CONUS, overseas, or deployed to combat

Formation: A group of Marines, usually of a particular unit or units (e.g., within a battalion), standing together, covered and aligned, Marines drill, stand inspections, hump, PT, and receive information and awards.

Fraternization: Contained in Marine Corps Manual 1100.4, fraternization is defined, roughly, as those relationships between officer and enlisted that are "unduly familiar," do not "respect differences in grade or rank," "prejudice good order and discipline" or "discredit the Marine Corps." All such relationships are prohibited. Such activities also fall under Article 134 of the Uniform Code of Military Justice, making fraternization a crime in extreme cases.

Full ride: A college or university scholarship paying 100% of a student's expenses while attending college

FOB (Forward Operating Base): Any secured forward position used to support tactical military operations, formerly considered behind the "front line." Presently, however, there is no "front line" in Iraq and Afghanistan, and a FOB and even positions behind it may indeed be considered the "front line."

Garrison: Life lived in a barracks or base environment rather than in the field or deployed. Generally being "at home" on the base to which one is assigned.

Gas chamber: A training evolution in which Marine recruits are acquainted with the reality that an enemy force, or even an individual, may use noxious and deadly gas or other nuclear or biological agents against him/her. Marine recruits are instructed in the use of gas masks beforehand, and must enter an enclosed chamber filled with non-lethal CS gas, chlorobenzylidene malonitrile. Recruits must enter, completely remove their masks, and then don and clear the masks to resume breathing air filtered by their mask.

Grunt: Affectionate term for a male Marine whose MOS is 03xx, infantry, although the designation is often extended to all male Marines in the "combat arms" MOSs, including a few within 02xx, Intelligence; 08xx, Field Artillery; one 13xx, Combat Engineers; and 18xx, Tanks and AAVs (Amphibious Assault Vehicles)

Gun squadron: A Marine Airwing unit whose mission is to deploy to combat

Gunny: A casual term for a Gunnery Sergeant (E-7)

Hail and Farewell: A social ceremony in which one Marine assumes a billet while another vacates it through PCS or retirement; often part of the "change of command" ceremony for upper echelon Marines

Head: A naval term used to identify what civilians call a "bathroom"

Head call: See above and imagine what activities transpire in "making a head call"

Hollywood Marines: Male Marines who attend recruit training at MCRD San Diego, California

Hootch: A tent, which is a Marine's house/home in the field or on combat deployments

Hump: A unit march or hike in full combat gear, usually many miles with a pack as well as an assigned weapon in addition to one's issued M16A2 rifle

IED (Improvised Explosive Device): Any improvised object created by a human being with the intention of maiming or killing another human being(s) using explosives

IOC (Infantry Officer Course): A 10-week course at Quantico, Virginia, trains/tests infantry officers through a wide array of leadership and tactical scenarios

Kevlar®: A composite material covered by fabric used as body armor because it protects, albeit limited, the human body against the assaults of small arms ammunition and shrapnel from explosions, mines, IEDs, etc.

Key Wives: Volunteers, usually Marines' wives, who attempt to educate families of service members through a network of information and activities, and who maintain contact with families left behind by Marines who have deployed

Kill Hat: Usually the third, or most novice, of a three-Marine team of Drill Instructors; her mission is to constantly unsettle, provoke, and berate recruits. The purpose of the "Kill Hat" and his/her behavior is to instill a constant vigilance against complacency in Marine recruits.

Land Nav (Land Navigation): A complex utilization of numerous tools to determine one's geographic location on the planet. Obvious "modern" tools are maps, protractors, lensatic compasses, but Marines proficient in the art of land nav utilize other primitive tools not subject to loss or destruction, such as pace counts, flora, assessing the skies and terrain, and numerous other intangible means of providing a very basic understanding of one's surroundings.

Leave: Scheduled time off used as the equivalent of a civilian's accrued vacation leave

Liberty: Time off between "normal" work assignments, such as evenings, roughly 1630 to 0730 not accounting for required formations or PT, weekends, and holidays that is not charged against a Marine's leave accrual

M16A2: The individual Marine's issued weapon for over 30 years, and often identified impulsively through rote memorization, as "...a light weight, magazine fed, air cooled, gas operated, shoulder fired weapon, capable of firing either semi-automatic or three-round burst," and some variation including all of the rifle's characteristics.

MARADMIN: "Marine Administrative Message" used to disseminate official information throughout the Marine Corps

MARC19: A grenade launcher

Marine Occupational Specialty (MOS): A Marine's actual job/occupation in the Marine Corps, after being a rifleman

Master Guns: A casual term used to address a Marine Corps Master Gunnery Sergeant (E-9). Not all Master Gunnery Sergeants appreciate this "affectionate" term, so caution is advised in its initial use toward a Marine of the E-9 rank.

Max: Used to designate the heaviest weight a Marine is allowed to be (e.g., "I'm at my max"), or to describe the best possible score to be achieved (e.g., "I max'ed the PFT.")

MCIs: Marine Corps Institute correspondence courses; some are required for Marines at various ranks and some garner college credit

MCMAP (Marine Corps Martial Arts Program): Emphasizes knowledge, physical fitness, and character exemplified through martial arts training/testing applicable to hand-to-hand combat; Marines may wear the earned colored belt in lieu of the prescribed khaki-colored web uniform belt.

MCRD (Marine Corps Recruit Depot): Initial training/testing sites for all male and female enlisted Marines. Females attend recruit training exclusively at MCRD Parris Island, South Carolina. With few exceptions, male Marines enlisted East of the Mississippi River also attend recruit training at Parris Island; male Marines enlisting West of the Mississippi attend recruit training at MCRD San Diego, California.

MCT (Marine Combat Training): Twenty-nine days of Marine Corps Common Combat Skills training, designed for all Marines assigned to non-combat MOSs, the mission of MCT is to further develop the basic warrior training recruits learned in Boot Camp to ensure that every Marine is, in fact, trained to be a rifleman, with the addition of additional basic weapons, before s/he enters the FMF

MECEP (Marine Corps Enlisted Commissioning Program): A program by which an enlisted active duty Marine is selected through an extremely competitive process to return to college while still on active duty, obtain an baccalaureate degree, attend OCS, and become a commissioned officer, a second lieutenant, in the United States Marine Corps. Some restrictions apply.

Medivac (Medical evacuation): The process of removing wounded or killed Marines from the battlefield or training environment

Mess chief: The highest ranking enlisted Marine in the chow hall; reports to the Food Service Officer (FSO)

Mess duty: A temporary duty assignment during which a junior enlisted Marine reports to the chow hall to perform daily work there instead of in one's MOS; often assigned as a form of punishment

MEU (Marine Expeditionary Unit): The smallest Marine Corps MAGTF includes infantry, helicopters, logistics, and a command element, usually commanded by a colonel

Mickey Mouse ears: Hearing protection worn in a variety of MOSs in conditions where hearing damage can occur

Midshipman: A student at the Naval Academy

Military Entry Processing Station (MEPS): A joint-service organization, with currently 65 locations geographically disbursed across the U.S. that screens military applicants to ensure suitability for military service based upon the various services' requirements; from this location, enlisted Marines ship to Boot Camp

MIMMS (Marine Corps Integrated Maintenance Management System): Per MCO P4790.1B w/ch 1 and 2, "The MIMMS is a set of manual procedures by which the effective use of personnel, money, facilities, and materiel as applied to the maintenance of ground equipment is controlled."

Mission: A Marine's task or assignment. Also used specifically to describe the number of contracts a recruiter is expected to write. Thus, one might hear recruiters discussing the efforts and success or failure to "make mission."

MOPP (Mission Oriented Protective Posture) gear: protective clothing Marines don in the event of chemical, biological, radiation, or nuclear training/warfighting scenarios

MOS (Military Occupational Specialty): A Marine's assigned occupation, training for which may result from attending a formal training school or OJT, on-the-job-training

MRE (Meal Ready to Eat): Rations in a sealed plastic bag eaten in the absence of chow hall facilities

MSG (Marine Security Guard): A B-billet, or secondary MOS, during which a Marine stands guard at one of over 100 U.S. Embassies worldwide

NCO (Noncommissioned Officer): The ranks of Corporal (E-4) and Sergeant (E-5) and the first rank, corporal, to wear the blood stripe and execute the sword manual. See Appendix A.

Obstacle course: A multi-station, cross-country course designed to challenge a Marine's physical and mental dexterity

Officer corps: Ranks from second lieutenant (O-1) through General (O-10). See Appendix A.

OIC (Officer in Charge): An officer responsible for the conduct and operations of a given unit

Ordnance: Ammunition, explosives of numerous varieties, and firearms

Officer Candidate School (OCS): Located at Quantico, Virginia, the venue where all Marine Corps officers attend initial training/testing. Although not integrated into training platoons, males and females train separately by sex but side-by-side, each witnessing daily the fact that the other sex performs the exact same training as their own

OpsO (Operations Officer): A member of the officer corps charged with ensuring that all research, planning, and execution of operations and training is conducted to the highest standard

Oxfords: Leather shoes requiring stripping, staining, and polishing, now replaced by Corfams, which do not require the same level of maintenance as leather shoes

Page 11: Location in a Marine's Service Record Book (SRB) where official counselings and disciplinary actions are logged

PC (Politically Correct): Overt attempts to eliminate the possibility of causing offense to someone, especially in respect to their sex, gender, race, or religion

PCS (Permanent Change of Station): Travel performed under orders from one duty assignment to another

PME (Professional Military Education): Educational activities required of Marines to be successful within their rank and to ensure their promotability to the next rank

Police: To "police" something is to clean it up, to square it away. Thus, to "police one's self" is to ensure that one is not lacking in physical presence (uniform), bearing, knowledge, or any other Marine Corps requirement.

Police call: A sweep of an area to pick up surface trash, especially cigarette butts

PFT (Physical Fitness Test): Evolved significantly over time to provide training standards and then test and ensure an appropriate level of "fitness" for all Marines with allowances made for sex and age

PT (Physical Training): Not necessarily including the PFT, usually performed as a unit numerous times per week. PT can include humps of a variety of lengths with a variety of pack weights and weapons, runs longer than the mandated three-mile PFT run, and a variety of other physically challenging training exercises

Quota: A term to describe a Drill Instructors "time off" from training recruits during which he/she performs duties in a support billet aboard the Depot; also used to describe a real or imagined Manpower requirement for a certain number of females to be in a certain MOS, a certain rank, etc.

RAINN (Rape, Abuse & Incest National Network): The U.S.'s largest anti-sexual assault organization

Rank: A Marine's achieved place within the overall Marine Corps rank structure. See Appendix A for the Marine Corps rank structure.

Red-patches: Sewn onto the front of the cover and on the outside of each leg of the utility trousers to identify landing support Marines; often confusing for the uninitiated, the patches do not identify grunts, but the next best things

Recruit: The designation of an individual in initial training (Boot Camp) seeking to become part of the enlisted ranks of the Marine Corps

Rifle Qualification (also called "Rifle Range," "Rifle Qual"): A Marine is required to shoot on a "known distance" course and a "field fire", with three possible awards at stake: Marksman, the lowest, Sharpshooter, or Expert, the highest, based upon the scores achieved on qualification day. Marines re-qualify every year.

RO (Reviewing Officer): The officer who attests that the markings on a Marine's performance evaluation, the Fitness Report (FitRep), are accurate and submitted in a timely manner; the RO is the commissioned officer who supervises the Reporting Senior (RS) on the Fire.

Road guard: during a formation movement such as a PT run or a hump, selected individuals from a platoon will break ranks to position themselves at intersections to stop traffic while the formation crosses.

ROTC (Reserve Officers' Training Corps): Programs existing at various levels of academia to provide military training of student volunteers some seeking a commission in the U.S. military

RS (Recruiting Station): RS is followed by the city name; site out of which Marine Corps recruiters work; location of initial administrative entry into the USMC by enlisted personnel through informational meetings, document preparation, and ASVAB and physical fitness training/testing

RS (Reporting Senior): The commissioned officer who supervises the Marine Reported On (MRO) responsible for rating that Marine's performance, conduct, and suitability for continued service

RTO (Recruit Training Order): Depot Order P1513.6A of 5 Dec 2005 governs recruit training at Parris Island, South Carolina, sets forth the policy for the conduct of all Marines, sailors, and government employees dealing with recruits.

Sandbagger: A Marine slacking off, not pulling her/his weight

SAW: M249 Squad Automatic Weapon

Sateens: Marine Corps olive drab utility uniform issued prior to the adoption of the camouflage utility uniform

Seabag: The equivalent of civilian "luggage," basically an olive drab-colored/green bag about the size of an adult human in which a Marine transports his or her military issue uniforms and other necessities.

Seabag drag: A disreputable display when a recruit or a Marine must pack up his or her belongings and leave the unit. Of note is that many Marines pack their seabag and move on in a positive sense, but doing the "drag" implies that one has been dismissed and is humiliated.

Seabees: The Navy equivalent of Marine combat engineers

Sergeant's Course: A course for Marine Corps sergeants to strengthen their leadership skills

Senior Drill Instructor (SDI): Highest ranking enlisted Drill Instructor training a platoon of recruits

Seps (Separations Platoon): Platoon of recruits waiting to return from initial training to the civilian world due to being dropped/discharged for administrative or medical reasons

Series Commander: A member of the officer corps, male for male recruit platoons and female for female platoons, who supervises and is responsible for the conduct of Drill Instructors and recruit training

Service Record Book (SRB): The "personnel file" of every individual Marine

Sexual harassment: Defined in Marine Corps Order 1000.9A of 30 May 06

Shelter-half: One sheet of canvas that is one-half of a two-person tent with snaps to connect to your tent-mate's shelter half for a whole tent

Shitbird: A consistently poor performer, not due to inadequacy or legitimate injury, but due to lack of commitment or subversive behavior. Every Marine has a person in her/his mind that personifies this label.

Shop: A Marine's worksite, whether an office or trade shop

Silent Drill Team: The USMC's 24-man elite rifle platoon that ceremonially performs at high-level national events; these Marines are stationed at Marine Barracks, Washington, DC, also known by its geographical location of 8th & I Streets.

SNCO (Staff Noncommissioned Officer): The ranks of Staff Sergeant (E-6) through SgtMaj of the Marine Corps (E-9). See Appendix A.

Squared away or "to square away": The state of being impeccable in terms of one's uniform, bearing, knowledge, job skills, or other aspects of being a Marine. To that end, also used as a verb to become "squared away" by fixing aspects of one's self. Also used to describe cleaning up or organizing a physical space, for example, to square away one's BEQ room for inspection.

Staff NCO Academy: PME resident school attended by SNCOs, and sometimes sergeants, to brush up on training and leadership skills, especially close order drill

Stateside: A location in the continental United States (CONUS)

Star Wars: A collection of scientific beliefs about the future of "modern warfare," including the belief that warfare would no longer be fought physically, on the ground, with humans [male warriors] pitted against each other like in the past. Although aspects of this way of thinking are correct such as "Fire and Forget," battles are still fought man-to-man and en masse, in guerilla and terrorist engagements

Squadbay: An open room to contain numerous racks (beds) and wall lockers in which Marines live

Swim Qualification (also, Swim Qual, "the pool"): All Marines must successfully pass swim qualification performed in the utility uniform, not in swimsuits; re-qualification is conducted based upon the level of proficiency achieved in the previous attempt

Tailhook Association: An organization that includes Naval and Marine Corps aviators, focusing on sea-based aviation and aircraft carriers, hence the name "tailhook" which is the hook attached to the bottom of the aircraft that catches the arresting wire suspended across the flight deck, stopping the plane from plummeting into the sea on the other end of the carrier

Tailhook Scandal: In 1991, at the 35th Symposium of the Tailhook Association in Las Vegas, Nevada, the two-day conference degenerated into debauchery (probably not the first time), resulting in numerous claims of sexual assault and harassment from males as well as females. A witch hunt for the perpetrators of the sexual escapades ensued, resulting in a few high-level firings and resignations, one suicide, and the investigation ultimately being stymied when the Navy and Marine Corps personnel in attendance at the convention closed ranks and refused to rat out their brethren

Tape: Measuring tape used to determine body mass for Marines bordering or on weight control

TBS (The Basic School): Located at MCB Quantico, Virginia, 26 weeks of actual officer training completed before an officer continues to his/her MOS school; very infantry-intensive

The Palms: Marine Corps Base Twenty-nine Palms, California

TIC (Troop in Combat): Marines, usually grunts but perhaps a convoy, on the ground requiring air support

Top: A casual term used to address a Marine Corps Master Sergeant (E-8). Not all Master Sergeants appreciate this "affectionate" term, so caution is advised in its initial use toward a Marine of the E-8 rank.

UA (Unauthorized Absence): A Marine is absent from his/her appointed place of duty without approval; Army equivalent is AWOL

Uniform Code of Military Justice (UCMJ): military law applicable to all members of the uniformed services

Vat cans: Containers used to transport temperature controlled chow to the field or training sites

Warrant Officer: Prior enlisted Marines who apply and are selected to remain as officers and specialists in their MOSs; warrant officers attend a separate The Basic School (TBS) before continuing to their officer MOS school

Weight Control: A program by which the Marine Corps attempts to assist Marines who have become overweight or are unable to pass the PFT because of weight issues

Wetdown: a celebration when someone gets promoted, usually involving consumption of and the wearing of obscene amounts of alcohol

WM: Abbreviation for "Woman Marine," a term used to differentiate females from males, usually in descriptive situations, such as, "Oh, she's a WM; she can't deploy with us." Use of this term is no longer acceptable.

Bibliography

Abbey, Antonio, et. al. "Alcohol, Misperception, and Sexual Assault: How and Why are They Linked?" In *Sex, Power, Conflict: Evolutionary and Feminist Perspectives*, edited by David M. Buss, 138-161. New York: Oxford University Press, 1996.

ALMAR 046/12. Unclassified. R 271120Z NOV 12. Subj: Change to the Physical Fitness Test (PFT). CMC Washington DC MCCDC C461TP/08AUG2008//.

Archer, Emerald M. "The Power of Gendered Stereotypes in the US Marine Corps." *Armed Forces & Society* doi: 10.1177/0095327X12446924 (May 2012): 359-391.

Baron, Larry, Murray A. Straus, and David Jaffee. "Legitimate Violence, Violent Attitudes, and Rape: A Test of the Cultural Spillover Theory." *Annals of the New York Academy of Sciences*, doi: 10.1111/j.1749-6632.1988.tb50853.x 1988, 528: 79–110.

Barnes, Julian E., and Dion Nissenbaum. "Combat Ban for Women to End." *The Wall Street Journal,* Politics and Policy, January 24, 2013, http://online.wsj.com/news/articles /SB10001424127887323539804578260123802564276.

Bellah, Robert N., ed., "The Dualism of Human Nature." In *Emile Durkheim: On Morality and Society: Selected Writings*. Chicago: University of Chicago Press, 1975.

Blau, Peter M. *Exchange and Power in Social Life.* New York: John Wiley & Sons, Inc., 1964.

Blomberg, S. Brock, Gregory D. Hess, and Yaron Raviv. "Where Have all the Heroes Gone?: A Rational-choice Perspective on Heroism." *Public Choice* 141 (2009): 509-522.

Boldry, Jennifer, and Wendy Wood. "Gender Stereotypes and the Evaluation of Men and Women in Military Training." *Journal of Social Issues* 57 no. 4 (Winter 2001): 689-705.

Brain, James L. "Sex, Incest, and Death: Initiation Rites Reconsidered." *Current Anthropology* 18 no. 2 (June 1977): 191-208.

Brown, Dee. *The Gentle Tamers: Women of the Old Wild West.* Lincoln: University of Nebraska Press, 1958.

Browne, Kingsley. *Co-ed Combat: The New Evidence That Women Shouldn't Fight the Nation's Wars.* New York: Sentinel, 2007.

Brownson, Connie, and Harold Dorton. "At the Margins of the Minors: Good Girls, Bad Girls, and Baseball Beyond the Big Leagues." *Popular Culture Review* 17 (2006): 67-78.

—. "The Battle for Equivalency: Female US Marines Discuss Sexuality, Physical Fitness, and Military Leadership," *Armed Forces & Society* doi:10.1177 /0095327X14523957 (6 March 2014).

—. "Rejecting Patriarchy for Equivalence in the U.S. Military: A Response to Anthony King's 'Women Warriors: Female Accession to Ground Combat'" *Armed Forces & Society* DOI: 10.1177/0095327X14547807 (27 August 2014).

Buss, David. *"The Evolution of Desire: Strategies of Human Mating.* New York: Basic Books, 2003.

—. *The Murderer Next Door: Why the Mind is Designed to Kill.* New York: Penguin, 2005.

—. "Strategies of Human Mating." *Psychological Topics* 15 no. 2 (2006): 239-290.

"California Lawmakers Pass 'Yes means Yes' Campus Sexual-assault Bill." *Los Angeles Times*, Local/L.A. Now, August 29, 2014, http://www.latimes. com/local/lanow/la-me-ln-california-yes-means-yes-sexual-assault-bill-20140829-story.html.

Carr, C. Lynn. "Tomboy Resistance and Conformity: Agency in Social Psychological Gender

Theory." *Gender and Society* 12 no. 5 (October 1988): 528-553.

Catechism of the Catholic Church. New York: Doubleday, 1994.

Ceccato, Vania, Robert Haining and Paola Signoretta. "The Nature of Rape Places." *Journal of*

Environmental Psychology 40 (2014): 97-107.

Chang, Johannes Han-Yin. "Mead's Theory of Emergence as a Framework for Multilevel Sociological Inquiry." *Symbolic Interaction* 27, no. 3 (Summer 2004): 405-427.

Chapkis, Wendy. *Live Sex Acts: Women Performing Erotic Labor.* New York: Routledge, 1997.

Clinton, Catherine. *The Plantation Mistress.* New York: Random House, 1982.

Cohen, David S. "Title IX: Beyond Equal Protection." *Harvard Journal of Law and Gender* 28, no. 2 (2005): 217-283.

Cohn, Carol. "'How Can She Claim Equal Rights When She Doesn't Have to Do as Many Push-Ups as I Do?': The Framing of Men's Opposition to Women's Equality in the Military." *Men and Masculinities* 3 no. 2 doi: 10.1177/1097184X00003002001 (October 2000): 131-151.

—. "Feminist Peacekeeping." in *The Women's Review of Books* 21, no. 5 Women, War and Peace (February 2004).

—. "Women and Wars: Toward a Conceptual Framework." In *Women and Wars*, edited by Carol Cohn, 1-35. Cambridge, UK: Polity Press, 2013.

Coker, Christopher. "Humanising Warfare, or Why Van Creveld May be Missing the 'Big Picture.'" *Millennium – Journal of International Studies* doi: 10.1177/03058298000290020201 (June 2000): 449-460.

Crosbie, Paul V. "The Effects of Sex and Size on Status Ranking." *Social Psychology Quarterly* 42 no. 4 (December 1979): 340-354.

Culver, Elizabeth M. "Women in the Service," *Annals of the American Academy of Political and Social Science* 229, The American Family in World War II (September 1943).

de Beauvoir, Simone. "Woman as Other." In *Social Theory: The Multicultural & Classic Readings*, edited by Charles Lemert, 367-370. Boulder, CO: Westview Press, 1993.

Demographics of Active Duty U.S. Military, Source: Defense Manpower Research. Date Verified: November 23, 2013: http://www.statisticbrain.com/demographics-of-active-duty-u-s-military/.

Dietz, Thomas, Tom R. Burns, and Federick H. Buttel. "Evolutionary Theory in Sociology: An Examination of Current Thinking." *Sociological Forum* 5 no. 2 (1990): 155-171

Dowling, Colette. *The Frailty Myth: Redefining the Physical Potential of Women and Girls.* New York, NY: Random House, Inc., 2001.

Du Bois, W. E. B. "The Spirit of Modern Europe." In *Social Theory: The Multicultural & Classic Readings*, ed. Charles Lemert, 183-186. Boulder, CO: Westview Press, 1993.

Dunn, Scott. "Experimental Combat Unit to Test Integrating Female Marines." The Official Website of the United States Marine Corps, http://www.hqmc.marines.mil/News/NewsArticleDisplay/tabid/3488/Article/160476/experimental-combat-unit-to-test-integrating-female-marines.aspx.

Durkheim, Emile. "The Dualism of Human Nature." In *Emile Durkheim: On Morality and Society.* Chicago: University of Chicago Press, 1973.

—. "Sociology and Social Facts," In *Social Theory: The Multicultural & Classic Readings*, ed. Charles Lemert, 79-82. Boulder, CO: Westview Press, 1993.

Easter, William. "The Marine Corps PFT: Not Equal, Not Fair." unclassified, submitted to USMC, Command and Staff College, 20 February 2009.

Edwards, Tim. "Queer Fears: Against the Cultural Turn." *Sexualities* (1998) vol. 1 (4): 471-484.

Elsesser, Kim M. "Does Gender Bias Against Female Leaders Persist? Quantitative and Qualitative Data from a Large-Scale Survey." *Human Relations* doi: 10.1177 /0018726711424323 (November 2011): 1555-1578.

"Female Troops in Iraq Exposed to Combat." *CNN.com*, World, June 28, 2005. http://www.cnn.com /2005/WORLD/meast /06/25/women.combat.

Francke, Linda Bird. *Ground Zero: The Gender Wars in the Military*. New York: Simon & Schuster, 1997.

Frye, Marilyn. *Willful Virgin: Essays in Feminism, 1976-1992*. Freedom, CA: The Crossing Press, 1992.

Gerth, Hans and C. Wright Mills, translated and edited. *From Max Weber: Essays in Sociology*. New York: Oxford University Press, 1946.

Glamour Daze: A Vintage Fashion and Beauty Archive. http://glamourdaze.com/history-of-makeup/1940s.

Goldberg, Steven. *The Inevitability of Patriarchy*. New York, NY: Morrow, 1973.

Gutmann, Stephanie. *The Kinder, Gentler Military: How Political Correctness Affects our Ability to Win Wars*. San Francisco: Encounter Books, 2000.

Hannoun, Antoine B., Anwar H. Nassar, Ihab M. Usta, Tony Zreik, and Antoine Abu Musa, "Effect of War on the Menstrual Cycle." *Obstetrics & Gynecology* 109, no. 4 (April 2007): 929-932.

Harris, G.L.A. *Living Legends and Full Agency: Implications of Repealing the Combat Exclusion Policy*. Boca Raton, FL: CRC Press, Taylor & Francis Group, 2015.

Herbert, Melissa S. *Camouflage Isn't Only for Combat*. New York: New York University

Press, 1998.

Hertz, Rosanna. "Guarding Against Women?: Responses of Military Men and their Wives to Gender Integration." *Journal of Contemporary Ethnography* 25 no. 2 (July 1996): 251-284.

Holm, Jeanne. *Women in the Military: An Unfinished Revolution*. Novato, CA: Presidio Press, 1982.

Huntington, Samuel P. "Officership as a Profession." In *War, Morality, and the Military Profession*, edited by Malham M. Wakin, 23-34. Boulder: Westview Press, 1986.

James, Alannah. "The Myth of Jessica Lynch: Gender, Ethnicity, and Neo-Imperialism in the War on Terror." *On Politics* 6, no. 1 (Spring 2012).

Jones, Sarah. "Sen. Gillibrand Takes on the Military Rape Culture and Gets Told She's Not an Expert," *Politicusa*, November 17, 2013: http://www.politicu-susa.com/2013/11/17/senator-kirsten-gillibrand-takes-aim-military-rape-culture-told-expert.html.

Kanter, Rosabeth Moss. "Some Effects of Proportions on Group Life: Skewed Sex Ratios and Responses to Token Women." *American Journal of Sociology* 82, no. 5 (March 1997): 965-990.

Kanuha, Valli Kalei, Patricia Erwin, and Ellen Pence. "Strange Bedfellows: Feminist Advocates and U.S. Marines Working to End Violence." *Affilia* doi: 10.1177/0886109904269053 (Winter 2004): 358-375.

Kolmerten, Carol A. *Women in Utopia: The Ideology of Gender in the American Owenite Communities.* Bloomington and Indianapolis: Indiana University Press, 1990.

Kaufmann, Walter, trans., *Beyond Good and Evil: Prelude to a Philosophy of the Future.* New York: Vintage Books, 1966.

Kier, Elizabeth. "Discrimination and Military Cohesion: An Organizational Perspective." In *Beyond Zero Tolerance: Discrimination in Military Culture*, edited by Mary Fainsod Katzenstein and Judith Perry, 25-52. Lanham, MD: Rowman & Littlefield Publishers, Inc., 1999.

King, Anthony C. "Women Warriors: Female Accession to Ground Combat." *Armed Forces & Society* doi: 10.1177/0095327X14532913 (May 2014): 1- 9.

Kümmel, Gerhard. "When Boy Meets Girl: The 'Feminization' of the Military: An Introduction Also to be Read as a Postscript." *Current Sociology* 50, no. 5 doi: 10.1177/0011392102050005002 (September 2002): 615-639.

Lee, Jesse, "The President Signs Repeal of 'Don't Ask Don't Tell: Out of Many, We are One." *The White House Blog* December 22, 2010: http://www.whitehouse.gov/blog/2010/12/22/president-signs-repeal-dont-ask-dont-tell-out-many-we-are-one.

Lerner, Max. "The Shame of the Profession." In *War, Morality, and the Military Profession* edited by Malham M. Wakin, 134-139. Boulder, CO: Westview Press, 1986.

Maginnis, Robert L. *Deadly Consequences: How Cowards are Pushing Women into Combat.* Washington DC: Regnery Publishing, Inc., 2013.

Manning, Lory. "Military Women: Who They Are, What They Do, and Why it Matters." *The Women's Review of Books* 21, no. 5 Women, War, and Peace (February 2004).

Marine Corps Order 1752.5, *Sexual Assault Prevention and Response Program*, 28 Sept 2004.

Marine Corps Order 6100.13, *Marine Corps Physical Fitness Program*, 1 August 2008.

Marine Corps Order P1020.34G w/Ch 1-4 of 31 Mar 03, *Marine Corps Uniform Regulations.*

"Marines Delay Female Fitness Plan after Half Fail Pull-up Test." *New York Daily News*, US, January 3, 2014: http://www.nydailynews.com/news/national/marines-delay-female-fitness-plan-fail-pull-up-test-article-1.1565216.

Martin, Daniel D. "Organizational Approaches to Shame: Avowal, Management, and Contestation." *The Sociological Quarterly* 41, no. 1 (2000): 125-150.

McDermott, John J. *The Drama of Possibility: Experience as Philosophy of Culture.* Fordham University Press, 2007.

Meštrović, Stjepan G. *Postemotional Society.* London: Sage Publications, 1997.

—. *The Trails of Abu Ghraib: An Expert Witness Account of Shame and Honor.* Boulder, CO: Paradigm Publishers, 2007.

Mitchell, Brian. *Women in the Military: Flirting with Disaster.* Washington, DC: Regnery Publishing, Inc., 1998.

Mittelstadt, Jennifer. "'The Army is a Service, Not a Job': Unionization, Employment, and the Meaning of Military Service in the Late-Twentieth Century United States." *International Labor and Working-Class History* 80 no. 1 (Fall 2011): 29-52.

Morris, Madeline. "In War and Peace: Incidence and Implications of Rape by Military Personnel." In *Beyond Zero Tolerance: Discrimination in Military Culture* edited by Mary Fainsod Katzenstein and Judith Reppy, 163-194. Oxford, England, 1999.

Moskos, Charles C. "Toward a Postmodern Military: The United States." In *The Postmodern Military: Armed Forces after the Cold War*, edited by Charles C. Moskos, John Allen Williams, and David R. Segal. New York, NY: Oxford University Press, 2000.

Myers, Sandra L. *Westering Women and the Frontier Experience 1800-1915.* Albuquerque: University of New Mexico Press, 1982.

Nepstad, Sharon Erickson. "The Continuing Relevance of Coser's Theory of Conflict." *Sociological Forum* 20, no. 2 (June 2005): 335-337.

Nietzsche, Friedrich. *Beyond Good and Evil: Prelude to a Philosophy of the Future.* Translated and edited by Marion Faber. New York: Oxford University Press, 1998.

—. *Twilight of the Idols, Or, How to Philosophize with a Hammer.* Translated by Duncan Large. New York: Oxford University Press, 1998.

Nix, Dayne E. "American Civil-Military Relations: Samuel P. Huntington and the Political Dimensions of Military Professionalism." *Naval War College Review* 65 no. 2 (Spring 2012).

Perilloux, Carin, Judith A. Easton, and David M. Buss. "The Misperception of Sexual Interest." *Psychological Science* 23 no. 2 doi: 10.1177/0956797611424162 (February 2012): 146-151.

"RAINN Urges White House Task Force to Overhaul Colleges' Treatment of Rape." Rape,

Abuse & Incest National Network (RAINN), NewsRoom, last accessed May 15, 2015: https://rainn.org/news-room/rainn-urges-white-house-task-force-to-overhaul-colleges-treatment-of-rape.

Ricks, Thomas E. *Making the Corps.* New York, NY: Touchstone, 1997.

Ridley, Matt. *The Red Queen: Sex and the Evolution of Human Nature.* New York: Penguin Putnam, 1993.

Rosen, Ruth. "The Invisible War Against Rape in the U.S. Military." *HNN History News Network*, March 24, 2014: http://hnn.us/article/155049.

Santangelo, Sage. "Fourteen Women Have Tried, and Failed, the Marines' Infantry Officer Course. Here's Why." *The Washington Post*, Opinion: http://www.washingtonpost.com/opinions/fourteen-women-have-tried-and-failed-the-marines-infantry-officer-course-heres-why/2014/03/28/24a83ea0-b145-11e3-a49e-76adc9210f19_story.html.

Schwartz, Eleanor Brantley and James J. Rago, Jr. "Beyond Tokenism: Women as True Corporate Peers." *Business Horizons* 16, no. 6 (December 1973): 69-76.

Seck, Hope Hodge. "Marines Delay Female Pullup Requirement Again, this Time Until the End of 2015." *Marine Corps Times* July 3, 2014: http://www.marinecorpstimes.com/article

/20140703/NEWS/307030068/Marines-delay-female-pullup-requirement-again-time-until-end-2015.

Segal, Mady Wechsler. "The Military and the Family as Greedy Institutions." *Armed Forces & Society* doi: 10.1177/0095327X8601300101 (Fall 1986): 9-38.

Sheppard, Leah D. and Karl Aquino. "Sisters at Arms: A Theory of Female Same-Sex Conflict and Its Problematization in Organizations." *Journal of Management* doi: 10.1177 /0149206314539348 (25 June 2014): 1-25.

Shields, Patricia M. "Sex Roles in the Military." In *The Military-More than Just a Job?*, edited by Charles C. Moskos and Frank R. Wood. McLean, VA: Pergamon-Brassey's International Defense Publishers, Inc., 1988.

Sjoberg, Laura. *Gender, War & Conflict.* Malden, MA: Polity, 2014.

Smith, Larry. *The Few and the Proud: Marine Corps Drill Instructors in Their Own Words.* New York, NY: W.W. Norton & Company, Inc., 2006.

Smith, Oliver Prince. "Retreat of the 20,000." *Time Magazine,* December 18, 1950.

Stiehm, Judith Hicks. "The Generations of U.S. Enlisted Women." *Signs* 11, no. 1 (Autumn 1985): 155-175.

Stremlow, Mary V. *A History of the Women Marines, 1946-1977*. Prepared for the History and Museums Division, Headquarters, U.S. Marine Corps. Washington, DC, 1986.

Studd, Michael V. "Sexual Harassment." In *Sex, Power, Conflict: Evolutionary and Feminist Perspectives*, edited by David M. Buss. New York: Oxford University Press, 1996.

Synnot, Anthony. "Shame and Glory: A Sociology of Hair." *The British Journal of Sociology* 38 no. 3 (1987): 381-413.

Titunik, Regina F. "The Myth of the Macho Military." *Polity* 40, no. 2 (April 2008): 137-163.

Tress, Bärbel, Gunther Tress, and Gary Frey, eds., "Defining Concepts and the Process of Knowledge Production in Integrative Research," 13-26, *From Landscape Research to Landscape Planning: Aspects of Integration, Education, and Application*. Wageningen UR Frontis Series (Book 12) The Netherlands: Springer, 2005.

Uniform Code of Military Justice (UCMJ): http://www.ucmj.us/.

United States Commission on Civil Rights. "2013 Statutory Enforcement Report: Sexual Assault in the Military." September 2013, Washington, DC.

United States Marine Corps Official Website > Home > Units > Recruit Training Regiment > 4th Recruit Training Battalion, Marine Corps Recruit Depot, Easter Recruiting Region, Parris Island, Couth, Carolina, http://www.mcrdpi.marines.mil/Units /RecruitTrainingRegiment/4thRecruitTrainingBattalion.aspx.

van Creveld, Martin. *Men, Women and War.* London: Cassell & Co., 2001.

VandeHei Jim and Chris Cillizza. "Bush Calls Kerry Remarks Insulting to U.S. Troops." *The Washington Post.* November 1, 2006, A Section, Page A08: http://www.washingtonpost.com/wp-dyn/content/article/2006/10/31/AR2006103100649.html.

Veblen, Thorstein. *The Theory of the Leisure Class.* New York: Penguin Books, U.S.A., Inc., 1899.

von Clausewitz, Carl. *On War*. Edited and translated by Michael Howard and Peter Paret. Princeton: Princeton University Press, 1976.

Wakin, Malham M. "The Ethics of Leadership II." In *War, Morality, and the Military Profession*. Edited by Malham M. Wakin, 200-216. Boulder: Westview Press, 1986.

Weber, Max. "The Types of Legitimate Domination." In *Social Theory: The Multicultural & Classic Readings*, edited by Charles Lemert, 122-125. Boulder, CO: Westview Press, 1993.

Weinstein, Jon. "Women's History Month: Leading Author in Women's Liberation Movement Continues Fight for Equality." *Time Warner Cable News NY1,*

March 18, 2014: http://www.ny1.com/content/news/178878/women-s-history-month--leading-author-in-women-s-liberation-movement-continues-fight-for-equality.

Weitz, Rose. "Women and Their Hair: Seeking Power through Resistance and Accommodation." *Gender and Society* 15 no. 5 (2001): 667-686.

Williams, Christine L. *Gender Differences at Work: Women and Men in Nontraditional Occupations.* Berkeley, CA: University of California Press, 1989.

Williams, John Allen. "Toward a Postmodern Military: The United States." In *The Postmodern Military: Armed Forces after the Cold War*, edited by Charles C. Moskos, John Allen Williams, and David R. Segal. New York, NY: Oxford University Press, 2000.

Williams, Kayla, with Michael E. Staub. *Love My Rifle More than You: Young and Female in the U.S. Army.* New York: W.W. Norton, 2005.

Winerip, Michael. "Revisiting the Military's Tailhook Scandal." *The New York Times*, Retro Report, May 13, 2013, http://www.nytimes.com/2013/05/13/booming/revisiting-the-militarys-tailhook-scandal-video.html?_r=0.

Yeager, Holly. "Soldiering Ahead." *The Wilson Quarterly* 31 no. 3 (Summer 2007): 54-62.

Yoder, Janice D. "Rethinking Tokenism: Looking Beyond Numbers." *Gender and Society* 5 no. 2 (June 1991): 178-192.

Zimmer, Lynn. "Tokenism and Women in the Workplace: The Limits of Gender-Neutral Theory." *Social Problems* 35 no. 1 (February 1988): 64-77.

About the Author

Connie Ann Brownson was born in Houston, Texas, on September 29, 1964, to Jack and Kathleen Pecha. She attended Mount Carmel High School in Houston, graduating in 1983. After three semesters at Texas A&M University in College Station, Ms. Brownson enlisted in the United States Marine Corps, completing Boot Camp at Parris Island, South Carolina, in March 1985. After completing training for MOS 1345, Engineer Equipment Operator, at Fort Leonard Wood, Missouri, Ms. Brownson began an 18-month tour of duty on Camp Foster in Okinawa, Japan in September 1985. Ms. Brownson completed her active duty career stationed at Camp Pendleton, California, and then New River Air Station in Jacksonville, North Carolina, completing her contract in January 1989 as a Corporal.

Returning to Texas as a Marine Corps Reservist after activation at Camp Lejeune, North Carolina, during Desert Storm/Desert Shield, Ms. Brownson graduated from Texas A&M in 1994 with a B.S. in Sociology and was inducted into Alpha Kappa Delta. She attended Officer Candidate School at Quantico, Virginia, in 1995, but was dropped due to a debilitating hip injury. She returned to Texas permanently as a civilian in March 1997 and entered the Master's program in Sociology at Texas State University. She received her M.A. in May 2003. Currently, Ms. Brownson is completing her Ph.D. in Geography at Texas State. She is also an Adjunct Instructor of Geography at Tarleton State University in Stephenville, Texas.

Standing duty at First Landing Support Battalion

In her spare time, Ms. Brownson conducts scholarly research and writes articles, contributing frequently to Armed Forces & Society. She enjoys her rural lifestyle, raising Boer goats on her small working ranch southwest of Austin, Texas. She enjoys competitive trail riding and participating in sporting clays tournaments. She is employed as an Administrative Assistant in Facilities Operations at Texas State University and has one daughter, Alexandra, one granddaughter, Charley Elizabeth, and another grandchild on the way.

www.ingramcontent.com/pod-product-compliance
Lightning Source LLC
Chambersburg PA
CBHW061716270326
41928CB00011B/2004